PERGAMON INTERNATIONAL LIBRARY
of Science, Technology, Engineering and Social Studies
The 1000-volume original paperback library in aid of education, industrial training and the enjoyment of leisure
Publisher: Robert Maxwell, M.C.

The Humane Use of Human Ideas

THE DISCOVERIES PROJECT AND ECO-TECHNOLOGY

Other Titles of Interest

AKASHI, H.
Control Science & Technology for the Progress of Society

BHALLA, A. S.
Towards Global Action for Appropriate Technology

DE GIORGIO, A.
Criteria for Selecting Appropriate Technologies Under Different Cultural, Technical and Social Conditions

ELLIS, T. M. R.
Control Problems and Devices in Manufacturing Technology

FORSLIN, J. *et al*
Automation and Industrial Workers Vol I Parts 1 and 2 and Vol 2

FORTI, A. AND BISOGNO, P.
Research and Human Needs

FRIEDRICHS, G. AND SCHAFF, A.
Microelectronics & Society: for Better or for Worse

HAWRYLYSHYN, B.
Road Maps to the Future

LASZLO, E.
Systems Science and World Order: Selected Studies

MAYOR, F.
Scientific Research and Social Goals

NAJIM, M. AND ABDEL FETTAH, Y. M.
System Approach for Development

PECCEI, A.
One Hundred Pages for the Future

A Related Journal

TECHNOLOGY IN SOCIETY*
An International Journal

Editors: G. Bugliarello and A. G. Schillinger,
Polytechnic Institute of New York

This interdisciplinary journal creates a single forum for the discussion of the political, economic and cultural roles of technology in society, social forces that shape technological decisions, and choices open to society in the use of technology. Subjects include, among others: technology assessment; science, technology, and society; management of technology; technology transfer; appropriate technology and economic development; ethical and value implications of science and technology; science and public policy; and technology forecasting.

*Free specimen copy available on request.

The Humane Use of Human Ideas

THE DISCOVERIES PROJECT AND ECO-TECHNOLOGY

Edited by

SHUHEI AIDA

Professor of Systems Science

University of Electro-Communications, Tokyo

on behalf of the

HONDA FOUNDATION

PERGAMON PRESS

OXFORD · NEW YORK · TORONTO · SYDNEY · PARIS · FRANKFURT

U.K.	Pergamon Press Ltd., Headington Hill Hall, Oxford OX3 0BW, England
U.S.A.	Pergamon Press Inc., Maxwell House, Fairview Park, Elmsford, New York 10523, U.S.A.
CANADA	Pergamon Press Canada Ltd., Suite 104, 150 Consumers Road, Willowdale, Ontario M2J 1P9, Canada
AUSTRALIA	Pergamon Press (Aust.) Pty. Ltd., P.O. Box 544, Potts Point, N.S.W. 2011, Australia
FRANCE	Pergamon Press SARL, 24 rue des Ecoles, 75240 Paris, Cedex 05, France
FEDERAL REPUBLIC OF GERMANY	Pergamon Press GmbH, Hammerweg 6, D-6242 Kronberg-Taunus, Federal Republic of Germany

First edition 1983

Library of Congress Cataloguing in Publication Data
Main entry under title:
The humane use of human ideas.
(Pergamon international library)
1. Technology and civilization—Congresses.
2. Science and civilization—Congresses. 3. Computers and civilization—Congresses. 4. Human ecology—Congresses. I. Aida, Shuhei, 1932-
II. Honda Zaidan.
CB478.H79 1982 303.4'83 82-11257

British Library Cataloguing in Publication Data
The humane use of human ideas.
1. Computers and civilization—Congresses
2. Electronic data processing—Congresses
I. Aida, Shuhei
303.4'834 QA76.9.C66

ISBN 0-08-027941-4 (Hardcover)
ISBN 0-08-027942-2 (Flexicover)

Printed and bound in Great Britain by A. Wheaton & Co., Exeter

Acknowledgements

Acknowledgement seems small recompense for valuable commentary and support. Nevertheless, I do wish to express my appreciation to a number of people who have participated in the DISCOVERIES symposia.

First, I am deeply grateful that I had the privilege to know and to be influenced by Dr. Soichiro Honda. My interest in sleep and dream research was kindled largely by the force of his personality, engineer's spirit and humanity.

I am especially indebted to the following chairpersons of the symposia in Tokyo, Rome, Paris and Stockholm: Professors R. I. Emori, E. R. Caianiello, J. C. Simon and G. Hambraeus. I take full responsibility for any omissions from this volume of the collective wisdom expressed by participants at the symposia.

I am also grateful to Mr. Bruce M. Adkins, rapporteur and consultant editor, formerly Chief of Public Relations and Information in OECD Nuclear Energy Agency, for his vital assistance in the preparation of this book and especially, besides writing Chapters 2 and 3, for translation of Chapter 7 from the original French manuscript and for substantial English revision of Chapters 1, 9, 10 and 16.

Much of the work presented in this book would not have been possible without the dedicated efforts of many people who worked and slept under extraordinary management conditions. In this respect, Messrs. S. Miyake and Y. Nakano have made significant and invaluable contributions.

I would like to thank Ambassador T. Shimoda, President of Honda Foundation, Messrs. H. Kawashima, T. Ueda, S. Shinomiya, M. Nishida and H. Yamamuro, Board Members of the Foundation who labored heroically to support the DISCOVERIES symposia financially as well as substantially.

Finally, I want to thank Miss M. Ueda, my secretary at the University of Electro-Communications, for her kind help in carefully typing the text for this book.

Shuhei Aida
Editor

Contents

CHAPTER 1

The Philosophy of "Discoveries": The First Discoveries Symposium, Tokyo, 1976

S. AIDA

Beyond the Culture Shock

The first DISCOVERIES International Symposium, planned and executed under the sponsorship of Dr. Soichiro Honda, was the beginning of a new scientific endeavour, "DISCOVERIES". The word, the plural form of an English word discovery, also stands for a rather lengthy phrase, "Definition and Identification Studies on Conveyance of Values, Effects and Risks in Environmental Synthesis".

Needless to say, modern civilisation was formed on the same foundation which gave birth to "philosophy", the source of human wisdom. Among others, "imagination", a human power for free conception, supplied the basis on which many types of civilisation were eventually formed. As sciences progressed in Europe, for instance, many false beliefs, which were then considered "scientific", were destroyed one after another. This is called the "science shock" of Europe in the Middle Ages. Europe thus entered the Dark Age which brought about the "witch hunt".

In tremendous agony at the gap between their thoughts and reality, human beings went through various stages before attaining the civilisation they have today.

Once again, however, civilisation is at stake today. In other words, human beings are experiencing another ''shock'' from the ongoing mechanisation of their civilisation.

''Mechanical civilisation'' was formed on two major undercurrents: ''Roman engineering'' and ''Arabian engineering''. Therefore, it is extremely important for us to know the difference between the two undercurrents as well as the patterns of their progress. Arabian engineering was, initially, the art of material refinery such as metallurgy. It developed into alchemy and then atomistic chemical theory. Thus, Arabian engineering was always the pursuit of material. On the other hand, Roman engineering began as civil engineering based on materials science. It developed into the art of measuring time and space and the integration of technologies which is, in contemporary terms, ''systematisation''.

The concept of ''time'' was something abstract in the agricultural civilisation that preceded mechanical civilisation. Time meant periodic cycles. People then did not measure time precisely; they felt, in such a cycle, the *flow of time*. As physical science made it possible to measure time precisely, the subsequent mechanical civilisation began to have a variety of far-reaching effects on society.

Thus, the Europe of the Middle Ages experienced the shock of many commonly-held thoughts being contradicted by developing scientific knowledge. Europe went through a series of such science shocks before it saw the age of Newton. For instance, Galileo Galilei advocated the heliocentric theory. Also, the traditional understanding of power and speed in connection with the movement of objects—the power and acceleration were in simple proportion—was ruled out.

In physical science-generated modern civilisation, where Newtonian determinism and probabilist thought and method coexist, its complexity has been intensified because of the amazing progress in computer technology. Take ''time'', for instance. It is now possible to measure physical phenomena that take place in a very broad range of time, from as long as thousands of millions of years to as short as one billionth of a second. In the field of the molecule, we now have molecular biology which even treats material and life at the same level.

Every bit of information obtained through research and development in many fields is fed into computers as "data" and, by the remarkably high performance of computers, such information is processed and used to probe all kinds of physical phenomena.

The Reinstatement of Spiritual Culture

As physical science brought about many discoveries and mechanical civilisation progressed, society found itself torn apart by an increasingly larger gap between spiritual culture and mechanical civilisation. In this dilemma, people were caught in ever increasing emptiness. They began to realise the importance of questioning substantive humanity.

All this happened because mechanical civilisation made such swift and sweeping progress. Countries aspiring to development introduced so-called "civilisation" regardless of their natural and racial features. One example was the uniform "Americanisation", so to speak, which posed various problems for the world's advanced industrial countries.

Perhaps it was the Japanese who opened up and imported technologies in the most "generous" manner, as they have done ever since the Meiji Restoration*. For them, what it meant was modernisation. Japanese intellectuals used to defend Japan saying she was progressive and full of vitality but, towards their

*The Meiji era: The Meiji era of Japan started with major political and social changes called the Meiji Restoration in 1868 and ended in 1912. During the period, the foundation of what later became a highly advanced capitalistic industrial state was built. The 1868 changes transformed the traditional feudalistic political system into a centralised governmental system. Japan under a new regime wiped out the vestiges of feudalism one after another and, simultaneously, carried out political and economical reforms. Japan looked for foreign systems and technologies to modernise herself and successfully transplanted them. In the economic field, state power intervened vigorously in industrialisation and, as a result, the country in a few decades accumulated very substantial industrial capital. By that time, however, advanced capitalist countries had entered the stage of imperialism. Japan's pursuit of imperialism thus resulted in the Japan–China war (1894–5) and the Japan–Russia war (1904–5). Japan further intensified her imperialistic and militaristic tendency which eventually led to her defeat in World War II (1939–45).

own people, they were pondering on the uniform introduction of mechanical civilisation.

During an Earth Science Association meeting in Dresden in 1886, Ogai Mori* engaged in a hot debate with E. Naumann†. Mori felt a sort of contempt in Naumann's speech on Japan and attempted to counter in the form of a table speech over tea.

Among others, Mori focused on what Naumann said of the pros and cons of Japan's motive for modernisation and its future. Naumann's speech conveyed the then widely held observation in Europe that Japan volunteered to open up to the Western civilisation it vigorously imported. In his criticism of Naumann, Mori argued that given foreign pressures Japan had to do so. The indiscriminate import of Western civilisation, Mori said, cast a shadow over Japan's future. In front of a Western audience, he criticised Japan for simply copying not only science and technology but also political and social institutions which had nothing to do with her own unique natural and racial characteristics.

In short, although he defended Japan's Westernisation to Naumann, Mori also told the Japanese about its "emptiness". He was entirely different from so many Japanese campaigners for "enlightenment" who, from that time to date, boasted to the West about Japan's "beautiful national characteristics" and to Japan about the benefits of "Bunmei Kaika" (civilisation). A self-avowed "foreign-educated conservative", Mori's attitude showed the concern of a man of letters about Japan which lasted all his life. This may seem especially interesting in view of the way Japan has since modernised herself. We must keep what Mori said in mind as the basic point of thought for the DISCOVERIES we are trying to internationalise. A little reflection will show that this basic thought is required to reinstate spiritual culture for human beings.

*Ogai Mori: 1862–1922—A doctor and writer, Mori was one of the outstanding intellectuals of Japan.

†E. Naumann: 1850–1929, during his stay in Japan from 1875 to 1885, helped modernise the country. He is also known as the founder of the archaeological "Naumann Elephant".

A New Viewpoint for Human Activities

After they achieved swift modernisation, advanced industrial countries found themselves bedevilled with the immense accumulation of scientific achievements and, in return, the emergence of public mistrust of technology.

In such circumstances, environmental issues have been spotlighted and, subsequently, technology assessment has been introduced. We should realise, however, that the US and other countries differ in cultural backgrounds and that technology assessment cannot be applied in a uniform fashion. The important thing is that each country develops her own response to unique environmental and cultural characteristics.

It was with this in mind that the first DISCOVERIES Symposium was held. The intention was not to confront existing problems and try to solve them. Rather was it to find out, from a whole new viewpoint of human activities, which problems would have to be solved in the future. The purpose was to seek, in these chaotic times, a philosophy for the prosperity of human beings. This might eventually lead to the ''from-art-to-technology'' transfer advocated by Dr. Honda.

What was the main objective of the first DISCOVERIES Symposium? It was to assess, from both hardware and software points of view, the modern mechanism; put human spirit into it, and revitalise it into a well-balanced one. For that purpose, a set of scenarios and strategies was to be prepared; for although people today feel some alienation from science and technology, they need a scientific probe to understand the interaction beteween people and cultural backgrounds, as well as the interactions between people and environment.

For instance, geneticists have suggested that even human thinking is determined by hereditary elements. That is, the genes determine not only the form of the brain but also the phenomena which take place in it: the thoughts. This scientific discovery should not frighten us. We should overcome such ''science shocks'' to search for the basis of a new civilisation. We have to separate ''honne'' (real intention) from ''tatemae'' (principle) so as to convey sound human wisdom to future generations.

This must mean an overall re-evaluation of the interaction between mechanical civilisation and individual culture. Specifically, we must examine and analyse the resistance to uniform Americanisation that has been developing in advanced industrial countries.

Background for western european science

Modern civilisation has always been the pursuit of science, West European-style science. For the purpose of correcting chaotic modern civilisation, therefore, it is necessary to know the basis of Western science and see what problems are built in. Thus we learn how those problems have arisen. At the DISCOVERIES meeting in Tokyo, Professor Eduardo Caianiello, then Director of the Laboratory of Cybernetics of the Italian National Research Council, spoke on *"Evolution and Revolution"*. In a reference to the Italian Renaissance, he said:

> The most dramatic example of cultural change is perhaps the transition, effected in less than two centuries, from the world of the Middle Ages, impregnated with Aristotelianism and religious prejudice, to that of modern science, open to free enquiry and centered on the Galilean method. It strongly supports the cybernetical view that both evolution and revolution are essential paradigms in the development of a complex system as human society certainly is.
>
> Leonardo da Vinci and Galileo Galilei's lives and struggles clearly illustrate the extremes to which the innovation necessary for the change to a new pattern of life may force mankind.
>
> A new revolution of comparable magnitude faces us today: the study of the past, made with the outlook of new science, may enable us to come out of it as masters rather than slaves.

Then, Professor Torgny Segerstedt, rector of Uppsala University, Sweden took up an interdisciplinary subject, "prediction", as the basic scientific approach. In a paper *"A*

Model of Interdisciplinary Research in Order to Make Prediction", he said:

> All prediction must be about the future situation of man in society. For that reason I believe we must start with a concept of society and its main characteristics. According to my theory we can discern in all societies three basic qualities or functions: (1) the reproductive function, (2) the socialisation function and the (3) productive function. These functions must be realised in all social groups (society is a group in sociological meaning) if the group is going to survive. This is a general sociological statement and it is valid for a nation as well as for an association of stamp collectors.
>
> In a society there may be different groups fulfilling different functions, as for example the family group around the reproductive function, schools and universities fulfilling the socialisation function, and working groups in factories or offices carrying out the productive function. In some societies the functions are carried out in the same group: that is the case in an agricultural society, which may be called a one-group society. In an industrial society there are three separate groups and the society can be called a three-group society.
>
> There is always an interaction between the basic function and the surrounding groups, that is, for example, between family and school, school and working group and working group and family. My main thesis is that every breakthrough in science must be of consequence for the functions of society. An interdisciplinary approach is to ask the question: what does, for example, a new discovery in biochemistry mean for family life? If we can predict such a consequence the next question will be: if we know that a discovery will influence the reproductive function what does it mean with regard to schools and work-groups? Or what does a technical innovation mean with regard to working groups, the educational group and finally for the family?
>
> The next step is to ask prominent scholars, as already has been practised by the Delphi method, what new discoveries they expect in their fields of research during the next three or four decades, and to try to determine what does that imply with regard to the basic

function. In such a way we may discover alternative futures and we will have a chance of making priorities with regard to our values.

I call such an approach scientific strategy (or strategic scientific play). I am sure that will cover all areas of mental and behavioral human activities.

Search for scientific thought

Whenever we try to use science and scientific thought as a viable methodology, we have to employ systems science.

The development of systems science owes much to space development. Because of this, however, systems science has been heavily weighted towards hardware, that is, the construction of engineering elements, rather than to the advance of "free human conception". On this, Professor Harold Linstone, of Portland State University, introduced in Tokyo a very clear approach to understanding in his paper *"Breaking the Chains of Traditional Systems Science"*:

> Systems science is historically the brainchild of science and technology. As such it has inherited the paradigms, the inquiring systems, the reductionist and analytic approaches so successful in the development of modern science. It is not surprising that our traditional education systems have a brainwashing effect as they compartmentalise knowledge. Even recent methodologies such as system dynamics, structural modelling, and technological forecasting reflect this strong bias.
>
> I consider it the foremost task of systems practitioners to consciously work toward a better balance—between analysis and synthesis, between reductionism and holism.* At least three of the five mental and behavioural activities have a strong holistic flavour: imagination, innovation and expression. The mention of

*Holism: The philosophical theory first formulated by Jan C. Smith that the determining factors in nature are wholes (as organisms) which are irreducible to the sum of their parts and that the evolution of the universe is the record of the activity and making of these wholes.

anthropology, art and culture in their description provides an important clue to this idea.

My talk examines the needed balance in two specific contexts: health and mobility. Both have been beneficiaries of powerful and sustained advances in technology. Medicine has conquered diseases and reduced the death rate, but we have paid scant attention to providing a reasonable quality of life and functional roles for the growing population of aged who represent the outcome of these medical triumphs. We have not provided a socio-cultural system to match the physical system.

Technology has shrunk the earth by incredible advances in the movement of people and goods (transportation) as well as ideas (communication). This is leading to rapid homogenisation of human cultures, to rootlessness of individuals, to acceleration of the discounting of both past and present, to a dramatic change in stimuli, and to creation of reality by media. We are barely beginning to recognise the impact of this technology on the human psyche.

Focusing on specific situations, we recognise a striking and urgent challenge to systems science: truly interdisciplinary—actually transdisciplinary—for innovation and invention.

Professor Toshiro Terano, of the Tokyo Institute of Technology, advocated "fuzzy" systems, a methodology of holism. He discussed the application of scientific thought in the imperfect real (as opposed to a perfect and unreal) world, and showed that much is to be expected from this "fuzzy" approach in future:

Individuals and Society: The mental and behavioral activities of human beings are founded on the preservative instinct of individuals and society. Their wish to improve circumstances is the potential of imagination, innovation and invention. Unfortunately, the survival policies of individuals are sometimes contrary to those of society. The development of communication and transportation never filled the gap in the society but aggravated it. If this trend continues, society must be disrupted.

One way to prevent it is the introduction of technological methodology to social problems. In this way we may find a compromise beween the subjectives and the objectives. However, there are defects in conventional technological methodology: in particular it avoids and disallows any vagueness in the objects. Human beings are so complex, flexible, subjective, multi-attributive and contradictive (in a word, "fuzzy") that such vagueness must be considered a natural attribute.

This author will talk about two kinds of methods which can be applied to such problems. One is a graphic model to probe the macroscopic structures of complex society, and the other is a "fuzzy" model which can deal with the mental activities of individuals.

Two phases in civilisation

When we analyse the fabric of modern civilisation, one useful way to do so is to examine hardware which has been rapidly developing. In urbanisation, for instance, we can see many examples of system engineering in material form. In Tokyo Dr. Wilfred Owen, senior fellow at the Brookings Institute, gave the following account of the world-famous "BART" transportation system in San Francisco—confirming that, as Professor Linstone pointed out, systems science is not always the powerful probing instrument it might and should be:

> *San Francisco's BART System*: San Francisco has completed its 118 km Bay Area Rapid Transit System (BART) which connects the city with outlying communities to the north and south, and eastward across the Bay to Oakland, Berkeley and other inland points. It is the first modern public transport system built in an American city.
>
> The project was conceived in the 1950s as an essential means of overcoming the congestion and pollution of automobile traffic and of assuring the vitality of San Francisco. The city, it was believed, could maintain and strengthen its position as the focal point of the region if good access could be achieved by a modern public transit system of sufficient quality to compete with the automobile. As the

project progressed it was promoted as a way to save energy, reduce highway requirements, create new growth centres, overcome suburban sprawl, and enhance the environment.

BART has not lived up to expectations. It has had effects that were not anticipated. Travel on the new trains is half the volume expected: most riders were not lured from their cars but are former bus riders; bus systems that once gave good service are now financially embarrassed; the system has not added appreciably to the centralising of functions in San Francisco, but has encouraged decentralisation; costs are so high that fares cover less than half the operating costs; the environment has not been improved nor has energy been saved; the volume of automobile traffic on the Bay Bridge has not been reduced; and a recent study shows that people who are carless have not been aided by this long-distance commuter-type rail line. The rich ride, and the poor pay (through increased property taxes and sales taxes).

But there are many unknowns about BART: its unmeasured effects on people and the region; the various public policies that might have been adopted to help make BART succeed; and the need for understanding how BART is part of the total transportation system, and how the transportation is part of the total urban system.

Modern civilisation is also supported by the "soft" systems which control human activities. These range from traffic control laws (which have much to do with urban dwellers) to old laws which have already lost force. Laws are both the newest ("legal cybernetics") and the oldest mechanisms to control human activities. From this viewpoint, Mr. Mario Miyahara, lawyer, discussed law in Tokyo:

Flexible Structure of Law: Defendant 'A', driving a car, tried to turn right at the central part of an intersection. At that point, a motorcyclist 'B', who was just behind 'A' tried to pass 'A' from the right. 'A' hit 'B', who was injured. 'A' was prosecuted and put on trial on a charge of misdemeanor in driving. The rules of traffic are stipulated in the Road Traffic Act. Of course, 'A' was not driving a car with those rules in mind. Should his driving happen to be in accordance with the rules, 'A' was only reacting to changing traffic situations. 'A' was not consciously obeying the rules.

Firstly, law provides rationalisation for whatever punishment a judge might give 'A' for misdemeanor. Law is a set of rules that judges and other public authorities employ for whatever happened.

Secondly, since law is an "expression", it is doomed to have a "fringe of vagueness" or it is doomed to be an "open texture". "Expression" has both "core of certainty" and "penumbra of doubt". As for the above case, the first and second trials found 'A' guilty. Contrary to lower rulings, however, the Supreme Court acquitted 'A' of a misdemeanor charge. This is the example of a creative judgement made by court.

Law's open structure entrusts courts with such an important power as "judicial lawmaking". Law is not just a strait jacket. Law has the kind of "flexible" structure that anti-quake buildings have. Thus, law can contribute to the progress and reform of society.

Civilisation assessment

Human activities intensify their complexity. If we try to see the relations between human activities and society, we need to take into account the spiritual, economic, ecological and environmental factors. In an attempt to apply technology assessment (an art developed in the US to generally assess such aspects) to civilisation, Professor Saburo Ichii, of Seikei University, discussed environment and the concept of "absurd pain":

> *The Environmental Problems and the Value Principle*: Modern civilisation has brought about pollution on a scale and of a nature never before expected. A set of transportation and safety-related problems is part of a larger pollution problem. Depending on the way transportation systems are developed, many difficult environmental conflicts will arise involving local inhabitants. As a philosopher, I think it necessary to work out a new principle of value as well as rules of application to deal with such conflicts.

As the civil society has developed, citizens have come up with various senses of right which have diversified and have conflicted with each other. Progress in transportation only helped such conflicts emerge on the surface and become more complex. In order to deal with this, it is necessary to work out value principles and rules of application—a "theory of social equality".

For a decade, I have been calling for less "absurd pain", that is pain inflicted on individuals for no reason. Many specific examples of this exist in the field of transportation, to which the value principle could be applied with benefit to all.

Professor Murray Turoff, of the New Jersey Institute of Technology (formerly on the staff of the White House) discussed the future impact of computers, communications and computer communications networks, new tools for the age of communication:

Assessing the Future Impact of Computers and Communications: The technology of these areas promises to continue its rapid rate of advancement with respect to significant increases of performance to price ratios. While there is little doubt "as to the magnitude of impact" these advances will have on society and such specific areas as energy and transportation, there is a great deal of uncertainty as to the manifestation of the impact and the exact nature of the resulting society.

Professor Turoff's presentation examined some of the potential futures that could result, and the methodological difficulties in assessing which would be the more likely directions taken. His paper examined basic changes that might occur in such areas as:

Decentralisation of Business Activities
The Paperless Office
Work at Home Concepts
Recreational Activities
Political Processes
Transportation Patterns
Employment Practices.

The impact which decision and planning processes could have on the nature of the above areas, and their ultimate impact on the individual and society, were explored.

Internationalisation and civilisation

As modern civilisation has developed, the progress of transportation and communications methods has led to world shrinkage in terms of time, resulting in rapid internationalisation. On the cultural front, however, international exchanges have inevitably tended to be superficial. Given such a background, the presentation of a culture must be made in such a way that people with different environmental, social and cultural backgrounds understand it. Thus it is necessary to develop presentation techniques that appeal to different people.

It was for this purpose that the Georges Pompidou National Centre for Art and Culture was built in Paris. It had been hoped that Mr. Sebastian Loste, who, as the senior administrative officer, had brought the centre from planning to execution, would be in Tokyo to explain the scheme. Unfortunately, Mr. Loste could not come to Japan, and instead his report was presented by Mr. T. de Beaucé, Cultural Counsellor at the French Embassy in Tokyo.

> *Reflections upon the Reasons for the Creation of the Centre Georges Pompidou*: In the first place, a project of this kind should reflect the current interests of a wide section of the public, taking account of generally felt needs. Without a broad outlook, such a Centre may soon degenerate into a club for 'insiders'.
>
> Secondly, future needs must be kept in mind. Not passing fashions, but the deeper trends of the time should be brought to the forefront. Public tastes and desires need to be anticipated, and an understanding of the present is impossible without at least some vision of the future, even a certain penetration into things to come.
>
> Finally, cultural matters stand in quite a different relationship to the past than does, for instance, progress in the sciences. The latter need feel little obligation to past effort; nor is a knowledge of the

history of a science needed to guide or inspire present efforts. It is true that the cultural heritage of the past is constantly re-examined and re-interpreted by succeeding ages, but it is never simply reproduced. Even artists who seem the most innovative have need, appearances to the contrary notwithstanding, to find in the past points of reference which shape and control their present inspiration. Even the most modern cultural undertaking cannot be divorced from the past.

The guiding principle of such an undertaking must, therefore, be an imaginative combination of the durable aspects of the present day, an anticipation of those of the future, and an acceptance of the legacy of the past in order to recreate the latter in terms more accessible to the present day.

The project, conceived and inspired by President Georges Pompidou, and put in train in November 1969, embodies the principle mentioned above.

All creations, whether living things or human endeavours, follow an inner logic of their own from conception to completion, shaped by an underlying necessity through the different stages of construction. Thus, there can be no truly fruitful outcome to either imaginative or inventive research unless a profound logical necessity guides the twists of imagination and the accidents of discovery.

Each age interprets somewhat differently the nature of this logical structure which is, of necessity, complex. It is by no means certain that future generations will take a different view of the proportion of invention and logic needed by such an undertaking.

But the fundamental notions are not likely to change: the joining together of State initiative with individual effort, the range of different disciplines, the making available to a wider public the value and benefits of the international heritage, illuminated by present notions and directd towards the expectations of the future.

Internationalisation also poses many problems for international politics. Even cultural exchange cannot be carried out without political considerations. Interdisciplinary study is required for the smooth prosecution of international politics. This was the basis of a presentation by Professor Masataka Kosaka, of Kyoto University:

> *Interdisciplinary Approach in the Field of International Political Science*: It is obvious that interdisciplinary studies are necessary in the field of international political science. I would like to list four examples:
>
> (1) First of all, the level and form of interactions among countries are remarkably diversified. This manifests itself in an inseparable interlocking of economic and political (in a narrow sense of the word) relationships. These interrelations cannot be understood in the context of traditional political relations. Nor is rational economic thought sufficient to do so.
>
> (2) The world today is culturally diversified. Whenever there is some sort of international relationship, the problem of cultural contact always arises. However, our notion of politico-economic 'common sense' is the product of our type of civilisation, rather than a universally-held concept. So, an anthropological approach is required.
>
> (3) Just like the first half of the twentieth century, its last half will be affected by technological innovation. Unless we understand it, we shall not be able to take a longer—more than medium range—view of international relations. Needless to say, since political intentions also intervene (as in resource issues), simple extrapolatory forecasts cannot bear much fruit.
>
> (4) In order to understand international relations, which are so diversified and have multiple "actors", massive amounts of data have to be processed. Of course, mere mathematical processing of such data is hardly sufficient to understand political phenomena. We need to know its basic requirements and limits, and then develop appropriate working methods.

The Tokyo Symposium, having heard the above papers, now turned to discussion in the hope of establishing "DISCOVERIES" as a scientific endeavour. From the general discussions, it was hoped to develop a four-part scenario as follows:

What has modern civilisation done?
What does it lack?
What is needed to revitalise it?
What can DISCOVERIES do?

However, the Symposium failed to produce any specific proposals and it was agreed to hold further "thought Olympics". Of course, it is not easy to make a successful appeal to world opinion. A mere desk theory will not be regarded as a serious appeal: experience has confirmed that this can come only through practice. For instance, a specific task of DISCOVERIES will be to convey the true "engineering spirit" to the people of the developing countries who are highly motivated for national construction. It is those people who can utilise fully what DISCOVERIES can offer: the enjoyment of and self-reflection on modern civilisation. To help them feel the real engineering spirit is what Dr. Honda called "the transfer of ideas from art to technology".

I am convinced DISCOVERIES should be able to show that such a spirit will remain for long as the "gene" of human beings.

CHAPTER 2

The Second Discoveries Symposium, Rome, 1977

B. ADKINS

THE Second International Symposium organised by the DISCOVERIES Project of the Honda Foundation took place from 7 to 11 November 1977 in Rome, under the auspices of the Italian National Research Council (CNR). The detailed arrangements were in the hands of a local Organising Committee chaired by Professor Eduardo Caianiello of Salerno University. There were some forty participants.

The Symposium was basically devoted to the relationships between cultures and society and the apparent crisis of credibility between science, technology and human wellbeing. In five sessions plus a final vigorous panel discussion with participation from the floor, the dilemmas of scientific and technological progress unmatched (it was generally agreed) by advancing humanity and developing human culture were analysed, explained and debated by pure scientists and mathematicians, philosophers and sociologists, technologists and medical doctors and others, representing disciplines which all too rarely work together or recognise their problems as parts of an interrelated whole.

Throughout the Symposium there was a clear consciousness that it would not be enough to order the world so that its foreseen stable-state population (estimated by the United Nations as 12,000 millions and expected to be reached about the middle of the twenty-second century AD) could be born, adequately fed and educated to survive in a more and more antlike community, only to disappear virtually without trace after 70 to 80 years of life. Besides discovering *how* this could be achieved without intervening disaster, it was fundamentally necessary to examine *why* it was a goal worth pursuing.

Thus, the Rome meeting concerned itself not only with science and technology, mathematical and philosophical techniques and methods, but essentially with moral, ethical and religious (in the widest sense of the word) aspects of life, culture and civilisation. That was a true interpretation of the "humane use of human beings" theme of the DISCOVERIES Project.

Science, Culture and Society

The first session was chaired by Father John Carrol, Dean of the Social Sciences Faculty of the Gregorian University in Rome. It was devoted to a lively but profound analysis of the structures of modern society as seen by an anthropologist, a structuralist, a planner and a Marxian historian. In the opening paper Professor Leo Apostel, of the University of Ghent, drew attention to the interactions between objective history (history which has happened) and the process by which it is recorded by historians. Professor Apostel maintained that the development of human society was an "autodynamic" process, evolutionary changes being basically due to internal rather than external causes. This meant that such changes were the results of internal conflicts within society, which must therefore be considered as an unstable system. Further, all human society was affected by environmental developments, resulting from technological developments, which meant that technology was a major driving force of history.

Because society must be continually renewing itself, its processes of reproduction must also affect its autodynamism. The resulting extremely complex interrelationships between societal structures and beliefs could show the need and justification for views of history changing as society itself developed. Professor Apostel, who professed himself an unorthodox Marxist, emphasised that the societal theories he had advanced were indeed Marxist—although not commonly so claimed by followers of "modern Marxism".

Professor Egeria Di Nallo (Bologna University) agreed with much of this, but pointed out that not all sociologists had the same view of society. To her, society was very complex indeed and any models adopted would be affected by the beliefs and social ambience of the modellers. Professor Di Nallo referred to the immense effects of capitalist consumerism, particularly on youth, and saw here the source of much of today's discontent which could and did lead to drug taking, terrorist activities and other evidence that youth felt deceived and denied the life which it had been promised by the preceding generation.

The Professor was particularly concerned with what she called *marginality*—the situation of a growing part of the population which was not required (either in the labour force or for material planning) to ensure the survival of the race. Such people were truly marginal, and their search for a *raison d'être* could readily lead to contestation. Yet who could deny them their concern for their own destiny?

The growing value and use of complex computer representations of developing society was described by Professor J. C. Simon (Paris University) who pointed out that modern computer programs and machines offered a hitherto undreamed of ability to carry out modelling of various societal structures. Nevertheless, he stressed a form of "uncertainty principle" which must always prevent any complex computer representation from being completely free from error. There was, he explained, a "wall of complexity" beyond which error susceptibility increased very rapidly indeed.*

Professor Piero Caldirola (Milan University), while not denying Professor Apostel's claim that societal ambience could affect attitudes of researchers, believed there were some basic truths of nature which were not so affected, and cited physics research in many countries as leading to the same truths whatever the political regime might be. In this Professor Caldirola was

*See full *Proceedings* of Symposium (Honda Foundation, Tokyo) pp 35–44.

strongly supported from the floor by Professor Robert Marshak of the City College of New York.

Professor Brunetto Chiarelli, University of Turin and Secretary General of the European Association of Anthropology, traced the story of mankind's gradual rise from purely animal status to his intellectual achievements of today. Pointing out that nearly all technological advance dated from the Industrial Revolution, he was not surprised that this sudden flowering of science had taken society aback. Scientists had suddenly reached the point where their social and ethical responsibilities went far beyond the simple creative impact of their work. There was never a greater need than now for bridge-building between these scientists, sociologists, and the people as a whole.

But Professor Chiarelli had an additional concern—and a reminder to "research workers who tend to forget that they belong to only one of the many living species on the surface of our planet . . . which is now, more than ever, being seriously modified by our own species". In this situation, he maintained, a new ethical attitude was imperative, leading—one might hope—to a new consciousness of the *interdependence* of all parts of world society, even including the marginal parts described by Professor Di Nallo. Although this involved some overtones of Marxism, it was essentially humanistic.

Science and Technology on Trial

Under the Chairmanship of Professor Robert Marshak, President and Professor of Physics at the City College of New York, the second session of the Rome Symposium was devoted to the "Golden Age" myth of Science and Technology, or more particularly to why so many intellectuals were today highly critical and mistrustful of advancing science and technology, maintaining that they had not—and probably would not—bring to the world the once promised Golden Age of peace and plenty.

The session was opened by Professor Harold Linstone (Portland State University, USA) who began by drawing attention to two extremes of current thought about the future of our world civilisation. On the one hand, very many intellectuals seemed convinced that world problems, ever more difficult to isolate and identify, were converging inexorably towards a single *megacrisis*. On the other hand, "most non-intellectuals are technological optimists: they are convinced of an affluent tomorrow for all mankind with the aid of technology".

Professor Linstone, who proclaimed his talk as a discussion of some navigation rules (for steering clear of the extremes cited) listed eight points to help an experienced navigator "skirt hazards skilfully, avoid catastrophes and make the journey smoothly".

His first point was that although continued exponential growth was admittedly and clearly impossible on our planet, the other extreme of no growth would mean an end to evolution and forward movement. "Searchers must therefore look for growth, but growth in a new dimension" he said, and in particular suggested intellectual or spiritual growth as valuable fields for exploration.

Secondly, recalling that increasing complexities in our world system mean that longer lead times and more knowledge are needed "to change course in a non-catastrophic way", Professor Linstone advocated forward planning that enlarged rather than reduced the options for future generations—concentrating on selected barriers to safeguard against misuse (for example leading to world destruction).

Other points included the encouragement of changes in individual attitudes ("the individual cannot escape responsibility"), encouragement of organisational innovation (the antithesis of over-bureaucratisation) and reduction of ignorance and "trained incompetence" ("those trained in science must develop a better understanding of behavioural and social systems; those trained in the so-called soft fields should be able to comprehend enough science and technology to discuss vital science policy issues"). Finally, and perhaps of special relevance to the DISCOVERIES Project itself, "do not stretch the paradigms of the scientific era beyond their realm of applicability".

In contrast to this wealth of navigational advice, other speakers in the session seemed rather to concentrate on what to do when the rudder fails and the rocks or sandbanks are already beneath the keel.

Professor A. Rasmuson, President of the Royal Institute of Stockholm and Professor S. Wikström of Stockholm University dealt with institutional problems which could lead (and had led) to questioning by the public of "scientific values", with widespread disappearance of confidence in practical scientists and even in the "scientific approach". The exploitation of scientific developments for profit—particularly in large industrial/commercial and often multi-national concerns—had led to calls for more government control and direction, yet this could result in unresolvable dilemmas when the exploiting concerns had much greater knowledge than the would-be controller of the science and technologies involved.

Reference was made to several recent disasters (notably the Seveso chemical disaster in northern Italy) where the full nature of risk from a large-scale industrial process was not perceived until that risk materialised as an accident. Official regulation was clearly needed, but the institution of appropriate regulatory criteria was a complex and difficult matter.*

The Seveso disaster was also looked at in discussion on a paper by Dr. Lucio Gaspari, Professor of Experimental Surgery at Chieti University, from which it was clear that medical doctors had for decades, probably for centuries, been obliged to come to terms with ethical questions that had only recently begun to seriously trouble scientists and technologists in other disciplines. Dr. Gaspari's remarks were in fact a most valuable commentary on some of the questions raised at the present time concerning many branches of science and the public: on information communication (when possibly the knowledge of approaching early death was better withheld from a patient); on the ethics of abortion, even when technically it was undoubtedly desirable; on relative values when lives—notably of "experimental" animals—were endangered or lost in order to preserve others; on

*cf. the "regulation" of nuclear energy.

the ethics of "buying and selling" medical attention and services which enabled wealthy individuals or important State personalities to obtain excessively expensive treatment (often with little or no real benefit) using finance and services which might well be more effectively applied elsewhere.

A further paper, presented by Professor Ulrich Briefs (Dusseldorf Institute for Economic and Social Studies) raised questions of ethical and social consequences of the exploitation of new scientific and technological developments: some of his points were emphasised by Professor Carlo Pucci (University of Florence) who in particular suggested that the complexity of social problems created by science and technology could grow much faster and acquire much greater significance than the science and technology concerned.

Although this session, like its predecessor, was characterised by much forthright declaration, it was difficult to perceive any major conclusions, as opposed to suggestions and recommendations from some speakers.

Returning to Professor Linstone, one such recommendation (under organisational innovations to be encouraged) was to "enforce equal justice (e.g. jail for political corruption and other white-collar crimes)". Although he did not specifically say so, it could be presumed that the "other white-collar crimes" should include corruption by scientists—particularly of their science.

Energy Versus Humankind

The third session of the Rome Symposium was chaired by Professor Harry E. Gunning, President of Alberta University, Canada. The session might well have had the wider title *Science versus Humankind*, for besides dealing with growing world demand for energy and how to meet it, it also examined whether the scientist was (or would be) responsible for major world disasters, and questions such as those raised by genetic engineering in relation to a continually expanding world population.

At the opening of the session, Professor Gunning in the chair drew attention to some frightening numbers which were already very much in the minds of most participants.

World population, he said, had taken:

50,000 years to reach 1 billion
100 years more to reach 2 billion
50 years more to reach 3 billion
15 years (1960–1975) to reach 4 billion.

Secondly, assuming population soon began to level off (a not particularly likely hypothesis but a useful one for practical forecasting), its total expected energy needs over the coming years would be some 10^{17}–10^{18} calories per year, to be set against estimated coal reserves of 8×10^{18} calories, and/or nuclear power reserves equivalent to 10^{20} calories without use of fast breeder reactors, or 10^{21}–10^{22} calories with breeders.

As to other sources, it was generally recognised that economically exploitable oil reserves (reserves which would not cost more in energy to extract than they would yield when extracted) were at most sufficient for some 20 years of present consumption rates, with natural gas in a similar position. Lastly, the so-called "renewable" sources (solar, wind, tides, geothermal, etc.) could not be expected to meet more than 1% or at most 2% of the world community's total demand for very many years (at least 100 and probably much longer).

This was the basic dilemma before the world, said Dr. Gunning, and for realists it must mean either that total world demand had to be drastically reduced, or that massive further use of coal and nuclear energy had to be envisaged. The decisions ahead would therefore inevitably be difficult and painful, and above all it would not be easy to convince some sectors of the population of their necessity.

This sombre message was taken up by Dr. Umberto Montalenti (Fiat Research, Turin), who presented data for the world and particularly for Italy which fully confirmed the "energy crisis" as real, challenging and probably unsurmountable without substantial changes in life styles at least in the more developed countries. Dr. Montalenti generally avoided discussion of alternative energy "mixes", but concentrated rather on two basic requirements:

to obtain more energy, by whatever means, and
to use less, by every possible economy.

In a special proposal for the latter, he introduced an interesting version of a "quasi-total-energy" system which he termed *energy cascading* and which envisaged energy, originally produced at high (1800°C) temperature passing through a cascade of stages where a portion was used and the remainder was passed on to the next (lower temperature) stage and so until finally the only rejected heat was at 5°C. But such a system required complex planning, including careful siting to enable each stage of the cascade to be efficient. It was however indicative of a new and spreading maxim in the energy supply field: do not use fuel which can produce high-grade (high-temperature) heat for producing only low-grade heat—e.g. for space warming. Instead, energy should wherever possible be used "at its most appropriate level".

The next speaker was Professor Luigi Mendia of Naples University, who countered Dr. Montalenti's proposals for optimisation of energy use from the *energy viewpoint* with a call for optimal use of technology in relation to respect and preservation of social freedom and dignity. Professor Mendia proposed that technical developments should be analysed as "pollution sources", geographic diagrams being constructed to show lines of *isopollution* joining points of equal pollution in the same way as isobars are used to join points of equal pressure. He showed that such isopollution lines were likely to coincide largely with the geographical concentration of technological developments. From this a strategy for pollution control, beginning with carefully planned land use and ending with necessary technical control measures, was built up.

Next, a representative of the Italian Montedison chemical concern, Dr. Di Giulio, gave an account of the chemical industry which was at once fascinating and sad. He explained how industrial chemists, once among the proudest servants of the community in providing fertilisers for improved crops, pesticides, and many other beneficial chemical products, were now reluctant to launch new products because their harmful effects, if any were found, would be given wide prominence and emphasis whilst their benefits would be forgotten. In the case of pesticides in particular, he envisaged discontinuance of the use of some of the more effective ones as being quickly followed by severe ravages of crops as the pests began to regain control. Though he quoted no figures, there was clearly a case for some form of cost/benefit and risk/benefit analysis.

This point was indirectly emphasised in a paper by Dr. Kazuhiko Atsumi (Tokyo University) entitled *Corroborative Harmony in Macroscopic and Microscopic Approaches towards understanding Human Beings*. Dr. Atsumi's plea that humans were in totality much more than the sum of their parts was a viewpoint with which no-one could disagree: this was confirmed by reports from two more medical doctors, Professor Siniscalco and Dr. Luzzatto from the Cornell Medical School in New York. Their two complementary papers dealt with the role of genes in the transmission of heredity and above all resistance to disease, and illustrated that neither genetic transmission could be considered separately from the environment, nor could (or should) important environmental changes be permitted without consideration of their possible genetic effects.

An incidental observation on the relation between genetics and race, dealt with at some length by Dr. Luzzatto, appeared to be that genetical classification could enable a very wide variety of diverse (and somewhat arbitrary) rules to be developed which could and would make the concept of race—beyond that of a single human race—virtually meaningless. The basic motivation here, in fact, would seem to depend on emotion rather than science.*

Finally Professor Giuseppe Cagliotti (Nuclear Engineering Institute, Milan Polytechnic) dealt with some applications of basic thermodynamic concepts (notably the 1st and 2nd laws) in fields

*For details see full *Proceedings* of Symposium, pp 184–5

beyond that of conventional thermodynamics. He showed, with great erudition, that in fact the entire societal ''milieu''—not excluding artistic activities such as music and poetry—could be analysed by a form of thermodynamic consideration in which *symmetry* and *information* are conflicting features, transition from one to the other being via *conservation* or *entropy* which affect quality but not quantity*.

Following the formal papers, Professor Piero Caldirola (Milan University) delivered a substantial declaration on the present and expected development of thermonuclear fusion—a timely reminder to all present that the energy situation today could one day (though probably not for several decades at least) be revolutionised by developments in applied physics.

This was perhaps the most practically-oriented session of the Symposium, and it was interesting to note that, despite some more or less superficial disagreements, the basic problems before mankind were recognised by all present irrespective of Scientific discipline (with captial *S*). It was also encouraging that none denied the basic value of Scientific approach and analysis in societal no less than technological system studies.

Communication and Information

The full title of the fourth session of the Rome Symposium was *The Intercultural Shock—tasks and failures of Mass-media Communication—Pollution through Information.* In fact its coverage went well beyond communication and information, and it proved to be a wide-ranging discussion of many facets of modern technology and their influence on ''the quality of life''. It was also noteworthy for bringing to a clearer focus than hitherto the somewhat different methods of reasoning found among the different disciplines represented—notably the ''pure'' physicists and the ''pure'' philosophers.†

*Again, reference to the full *Proceedings* (pp 191–7) is recommended.

†If the Symposium was noteworthy (as it certainly was) for bringing together Occidental and Oriental approaches to understanding, it was no less noteworthy for similarly combining many branches of academy which in normal life are separately and independently pursued.

The session was chaired by Professor Roger Revelle, Director of the Center on Population Studies at Harvard University, USA, who brought a degree of *optimism* to the meeting (see remarks on optimism by Professor Linstone in the second session) by recording that although poverty was still to be found in the technological society of the United States ". . . the share of the national income received by the richest 20% of the people is less than six times that of the poorest 20%". This he contrasted with—as an example—Kenya, where the income of the rich was eighteen times that of the poor, or Equador, where nearly three quarters of the national income went to the richest 20%, and only one fortieth to the poorest. "In all technically advanced societies" the Professor concluded, "the work of the poor is becoming less degrading . . . as could be seen from the virtual disappearance of servants in the US".*

A further point made by Professor Revelle, which is well-known but perhaps not often enough recognised, was that manual labour, however morally ennobling it might be counted, was an extremely expensive way of producing power. Considering only the minimum cost of maintaining a man on a very meagre diet—estimated as at least 15 US cents per day—and that the maximum work output obtainable from the man would at best be about one kilowatt-hour, this was obtainable from a diesel engine or electric motor for less than one fifth of that cost. That made any form of slavery highly uneconomic, whether it might be "conventional" slavery involving chains and cotton-picking, or the "brutalising drudgery" of the early days of the industrial revolution.

A particularly interesting paper in this session, and one which did in fact consider some of the problems of mass-communication, was that of Professor Barrington Nevitt (Toronto University) who examined the differences between the intellectual processes of communication in western and eastern countries. The western "presentation of information", as Professor Nevitt saw it, was in

*The provocative implication that service is necessarily degrading was—perhaps fortunately—not taken up in subsequent discussion.

visual space with "separate centres and fixed boundaries": these had to be examined (read) by a logical and sequential process which not only required *time* but was disturbed by incorrect sequencing. On the other hand, the eastern method comprised a *simultaneous presentation*, with "everything going on all at once and all the time": this was *acoustic space*, structured like a symphony and having "centres everywhere and boundaries nowhere". The comparison between conventional sequential reading and "speed-reading", where whole pages of type are—or should be—absorbed in one and the same instant, was stressed in the subsequent discussion whence it appeared that many of today's youth, brought up with a large and continuing input of televised information, had developed a facility for instantaneous comprehension of a *screen full of data*, and in fact were in difficulty if confronted by sequential information presentation as in a book or newspaper. According to Professor Nevitt, these two types of presentation ("diachronic" and "synchronic") were difficult if not impossible to translate mechanically one to the other—and if such interchange were possible it must be within the human brain.

Many conclusions were drawn in the paper from this and its dependent theses: probably the most stimulating result however was the emphasis on the "complementarity" of a communication system and information communicated, suggesting that ignoring the former would almost inevitably lead to ambiguity and inaccuracy in the latter.

Professor Maurice Marois, Director of the *Institut de la Vie* in Paris, derived further optimism both from the experience of his Institute and (particularly) from observed and recorded biological and anthropological evolution. Stressing the remarkable perseverance of life in ensuring its own survival*, he found in its "policy" of conquest, self-expression and adaptation a guarantee of "a great future ahead". He recalled that our planet earth is expected

*A single emission of male seminal fluid contains between 200 and 300 million spermatozoa, a number equal to the population of Western Europe, while the ovaries of one woman contain 700 thousand ovules, 400 of which are emitted at the rate of one every 28 days over the 30-year span of the woman's reproductive life.

to remain habitable for man for another six thousand million years, "so that humanity has so far covered only one ten-thousandth part of its journey since the first man appeared". But the Professor warned that mankind, now at the top of the evolutionary tree, had acquired ample means to cut himself off from destiny. It was to avoid this that man must profit from past and present experience in preparing for the future: it was here that information, communication and understanding were vital for intelligent survival. Straightforward exponential multiplication of population must rapidly lead to penury, scarcity and elementary limitations of space, which would impose more or less brutally the controls that intelligence had failed to provide.

Professor Marois made it clear that the objectives of the *Institut de la Vie*, in studying these and related matters, were in many ways similar to the objectives of the DISCOVERIES Project. Whether or not this was cause for rejoicing (he could have noted that various other institutions, groups, projects and programmes had equally high ideals and ambitions) must depend on whether theoretical science, philosophy and speculation could lead to practical actions—to the skilful navigation of the "smooth journey" advocated by Professor Linstone in the second session.

Next, a paper by Professor Gabriele Morello (Palermo University) dealt with the time-sequencing of information transmission over substantial periods from past through present to future, and analysed the dependence of any one of these periods on the others. Showing that different people had different "orientations" in this respect—interpreting received data according to their method of perception (in relation to the past, in relation to the present alone or in combination with past and even future, or in relation mainly with the future) Professor Morello used these orientations to arrive at tentative specifications for business executives, personnel managers, salesmen, researchers, and even successful partners in marriage.

Next Professor Toshiro Terano (Tokyo Institute of Technology) gave a descriptive talk on "Fuzziness in Language" and its implications for information communication. The use of "Fuzzy Set" symbolic techniques (whether these are truly mathematical or not is perhaps yet to be demonstrated) seems undoubtedly capable of helping certain forms of systems analysis, and Professor Terano's development of a somewhat "fuzzy" theme was instructive and illuminating. It certainly threw into relief some of the differences in the methodologies of comprehension to which Professor Nevitt had drawn attention.

Other contributions in this session were from Professor José Roberto Moreno (Las Palmas) and Professor José Garcia Santemases (Madrid) whose remarks certainly supported the Chairman's view—at the end of his paper—that "All experience shows that the solutions to one set of human problems almost always bring new problems in their train".

"We should not ask more from new technology" he said, "than that it is likely to make things go from bad to better".

In view of the frequent accusations that new technologies are in fact doing just the opposite, it may be opportune to remark that, in reality, it all depends on how they are used. It is up to us.

Global Planning

Under the chairmanship of Professor Zvonimir Damjanović, Director of the Centre for Multi-disciplinary Studies at Belgrade, Yugoslavia, the fifth session of the Rome Symposium was devoted to a combined oriental/occidental approach to the problems of forward planning in a world where the technologies of electronic computing, colour television, supersonic travel and space exploration exist alongside the ancient customs of some countries where superstition and magic still rule much of society—and where much of that society survives only in constant struggle against starvation and poverty.

The session was particularly noteworthy for two reasons. In the first place, there was a short paper from a representative of a Developing Country (Nigeria) which, with an elegant simplicity that must itself have inspired those who heard it, emphasised the enormous and sudden impact of the explosion of advanced technology on countries where—well within living memory—"everything was viewed in terms of magic and the intervention of ancestral spirits''. The paper was presented by Dr. Donald Ekong, Vice-Chancellor of the University of Port Harcourt, who also had a lesson for those who—comfortably installed amidst the material benefits of highly industrialised modern societies—advocated a return to ''nature and the simple life'' as the solution for all our problems. ''It is a myth not related to reality'' said Dr. Ekong ''to regard the pre-technology era as a kind of heaven-on-earth''.

The second specially significant contribution came from the other end of the development scale. A long and deeply philosophic paper from Professor Ilya Prigogine of the Free University of Brussels and winner of the 1976 Nobel Prize for Chemistry, described some of the latest mathematical methodologies for analysing non-linearities in chemical and other reactions at points close to loss of structural stability and the appearance of new structures (''bifurcation points'').

Although these methodologies originated in the so-called natural sciences, they were clearly applicable to other large, complex and randomly-coordinated (if this is not too paradoxical) systems such as those of human communities within the framework of overall society. Professor Prigogine spent some time examining in this way the impact of developing urbanisation following the implantation of a town in a hitherto mainly rural environment.*

Among other papers in this fifth session of the Symposium, one from Professor Vincenzo Caglioti (Rome University) drew attention to the need for practical application of theories in order (a) to verify by experiment their validity and (b) to derive some benefit for mankind from the work. This was perhaps a useful reminder to those with tendencies to allow their academic cogitations and conclusions to become an end in themselves, rather than means to an end.

*No summary (let alone a very short one) could possibly do justice to this introduction to ''guided intuition'' in approaching the present limits of mental representation and analysis. The reader is recommended to the full *Proceedings* of the Symposium, pp 297–311, or better still to Professor Prigogine's original paper (in French) which was distributed to participants.

Another paper, from Professor Gunnar Hoppe, Vice-Chancellor of Stockholm University, described some of the remarkable technological tools now used for "remote sensing" of data which were needed (e.g.) for practical management of agriculture or forestry on a worldwide, or at least continental, scale. Observational aircraft, satellites, etc. of immense complexity were now able to provide such data in a constant stream of dynamic information: combined with computers for analysis and trend-prediction, they opened for the world a new level of surveillance, verification and control.

Professor Jacques Della Riccia (Ben Gurion University, Israel), clearly another participant whose academic work was necessarily close to the ground of practical problems, reiterated a call that had been heard earlier in the Symposium, notably from Professor Marshak of New York City College. While not denying the fascination of new types of symbolic mathematics and logic, Professor Della Riccia reminded all his colleagues of the highly reputable ancestry of more conventional systems for representing dynamic processes, and insisted that these still had important—indeed vital—contributions to make to an improved world and society.

This view was supported by Professor Sidney Michaelson (Edinburgh University) who gave an account of some of the ways in which modern computers could make possible practically-oriented "modelling" operations in "societal engineering", which otherwise would be very much more intractable and even impossible.

However, the main theme of Professor Michaelson's intervention was that sociologists, economists, researchers of all kinds who were engaged in such modelling operations had yet to achieve any reasonable understanding of the society they tried to analyse. This lack of understanding applied no less—and probably even more—to the politicians and administrators who controlled most of our daily life and much of our destiny. The Professor's plea was therefore that, rather than spending increasing funds on the *technical machines* (the working tools) of societal analysis and control, much more should be devoted to basic human contacts and enquiries, to developing fundamental understanding without which the machines could be useless, or even harmful.

In an impassioned but perfectly reasoned appeal to all the erudite participants in the meeting, Professor Michaelson called especially for more attention to be given to the thoughts and attitudes of young people (those who in a few years would be running the world) towards the life they found about them and how they would hope to see it evolve.

Finally in this session, Mrs. Nanami Shiono, former researcher in social sciences in Japan, now living in and writing about Italy, contributed both oriental wisdom and feminine understanding to the Symposium. It was perhaps both fitting and relevant, at a meeting in this Eternal City which has contributed so much to the evolution of both civilisation *and religion* through the ages, that one clearly imbued with another form of these two attributes should make the point that the "eastern religions"—Confucianism, Shintoism and Buddhism—had never inspired "wars of conversion", as had many others including Christianity.* The reason, she said, was evident: it was *tolerance*, and recognition that no single religion must necessarily make all others invalid.

The same tolerance was needed in other affairs of mankind, including attitudes to work and to leisure. Mrs. Shiono had been surprised to find so much work regarded with distaste in western civilisations, and the substantial divisions between "leaders" and "workers". Both were essential, and in democratic society both should cooperate, for the general good, in "tolerance and understanding" of each other and each others' problems. In this way democratic society itself—which meant different things to different people—would also evolve for the benefit of all people.

Roma Locuta Est

A final (sixth) session of the Second DISCOVERIES

*G. B. Shaw once remarked "There is only one religion, though there are a hundred versions of it".

Symposium was devoted to a review of the five preceding sessions and a general discussion of results and conclusions. This sixth session, under the chairmanship of Professor Eduardo Caianiello, was honoured by the presence of Dr. Mario Pedini, Italian Minister for Science, and Professor Ernesto Quagliariello, President of the Italian National Research Council (CNR) under whose auspices the Symposium had been organised in Rome.

Despite the quite considerable divergence of views concerning some of the matters discussed earlier in the meeting, the concluding session at least demonstrated complete unanimity in believing (a) that the interdisciplinary approach of DISCOVERIES was not only valuable but essential for the resolution of today's problems of population, food and energy supply, raw materials, and above all improved human understanding; and (b) that the bringing together of Eastern and Western thought and cultural approaches in the unique way that the series of DISCOVERIES Symposia made possible comprised a novel and exceptional contribution to facing world problems on a world basis. There was also substantial agreement that, although novel methods of scientific approach (especially of symbolic representation and computer-based analysis) were to be encouraged, it would be very stupid indeed to allow these to displace the traditional methods of analytical deductive and inductive science which had served so well for centuries.

A marriage of these traditional methods with the new was both desirable and probably essential. As Professor Alwyn Scott of Wisconsin University declared, it was highly important to avoid any form of dogmatic limitation to the techniques used in addressing the problems of modern life and society. "The rationalist assertion", he said "that counting, measuring, logical inference, abstraction, generalisation and mathematical formulation provide the only sure approach to human knowledge rests on the intuition of certain scientists; while the emerging view that other modes of perception can be useful, and may even be essential, is supported by a rational argument".

This remark appeared to have been stimulated by a major (and somewhat paradoxical) duality of reasoning which had been noticeable throughout the meeting, detectable in particular in exchanges between ''pure'' physicists and ''pure'' philosophers (if either term is acceptable to those so described). For while both had paid continual tribute to ''the scientific method'' as the most promising way to a solution of the world's problems, clearly both had rather different concepts of what the scientific method was.

It was Professor Robert Marshak of New York City College who—as he had done several times previously—brought the discussion to *practical levels* by insisting that the real test of any approach, scientific, intuitive or other, was ''whether it worked''. It was Professor Caianiello who, likening much scientific method and belief to religious method and belief, insisted that neither—even when providing the answers required—should be considered exclusive of all others.

The problems of the world, said Professor Caianiello, were too great, too complicated and too universal in content to be resolved without recourse to every intellectual approach that could be found. Pure mathematics, physics and chemistry and medicine should combine with philosophy, social science and, above all, the humanities. The Future was the concern of all the world and all mankind: its fruitful pursuit demanded a holistic approach at every level, scientific, technological, artistic and humane.

The conclusion—simple of expression but most complex and difficult of realisation—was aptly expressed by Professor Barrington Nevitt of Toronto University. Calling for a much closer collaboration between the ''artistic'' and the ''numeric'' approaches to Science, he advocated that ''engineers should be encouraged to become poets, and poets to become engineers''. All between must inevitably benefit.

CHAPTER 3

A Communications Society: The Third Discoveries Symposium, Paris, 1978

B. ADKINS

Introduction

The Third International Symposium organised by the DISCOVERIES Project of the Honda Foundation took place from 23 to 27 October 1978 in Paris, under the high patronage of Professor Pierre Aigrain, French Secretary of State for Scientific Research. The detailed arrangements were in the hands of a Scientific Committee under the Chairmanship of Professor Jean-Claude Simon, of the Pierre and Marie Curie University in Paris. There were some fifty participants.

In devoting the Symposium to "Communication in Human Activity", the organisers had clearly put a finger on one of the most important, and most controversial, facets of today's world. As Professor Aigrain said in his opening address, instant "real time" electronic communication throughout the world had given the human race a tool which so far surpassed older inventions based on the written or printed word that it could lead to unprecedented advances of civilisation and humanity. But also, and like so many other developments of science, if misused or abused it could very easily lead to our destruction.

Dr. Takeso Shimoda, President of the Honda Foundation, also emphasised the increasing problems of humanity in controlling its inventions. Citing the (then recent) accidental disintegration of a

Russian nuclear-powered satellite over Canada, leading to fears of radioactive pollution in that country, he called for general recognition of the entire world as a "single village", in which no activity in one area could be without its influences in others. These influences, said Dr. Shimoda, could only be understood through a full exchange of views and ideas from all the intellectual disciplines involved, which was one of the basic objectives of the DISCOVERIES Project.

Professor Eduardo Caianiello of Salerno University, who had played a leading role in the development of the DISCOVERIES programme from its beginning in Tokyo nearly three years previously, paid a special tribute to those objectives. During recent visits to many parts of the world, both East and West, he had found everywhere a strong feeling of impending crisis, an urgent desire to meet this with appropriate action, but a widespread lack of certainty as to what such action should be. This was surely where DISCOVERIES should try to help, and the fact that the Project's Symposia were bringing together the most advanced realms of science and technology, as well as philosophies of the past and present throughout the world, emphasised what Professor Caianiello believed was an inescapable truth. It was not the case, he said, as some maintained, that the evils brought about by the technologies of civilisation could be overcome without further resort to those technologies. All technologies, in fact, were neutral—only the use made of them depended on the will of man.

Professor Jean-Claude Simon, introducing some of the subjects to be discussed during the Paris Symposium, emphasised the complete freedom of speakers and other contributors to say what they thought and believed, without the constraints of "official policies" whether of their governments or their learned institutions. He had been particularly impressed by this during the previous meeting in Rome, and was sure the same liberty would be a feature of the new meeting now begun. We had reached the point, he said, where nearly 50% of all working occupations were concerned with the production, handling (processing) or communication of information. This was perhaps the greatest growth industry of our time, with immense technological, economic, and above all societal implications. That was why it was essential to examine the situation with the most profound honesty and concern for the present and future well-being of the world.

Communications Society Defined

The first working session of the Paris Symposium was devoted to defining and describing the type of society subsequently to be discussed. It began (as did all the working sessions) with two main "scene-setting" papers, which were followed by a number of shorter presentations and general discussion in the form of "round tables" with participation from the floor.

The two main papers were given by Professors Alain Touraine and Louis Dumont, of the French *Ecole des Hautes Etudes en Sciences Sociales*, who set out to describe the historical and current sociological processes leading to a totally interdependent world-wide, communications-based "programmed society". Such a society was presented (by Professor Touraine) as the most preferable of three possible developments, the others being *decadence* (as evidenced by preference for immediate consumerism rather than preparation for the future), and the evolution of various *counter-cultural movements* implying changed values (preoccupation with more "primitive" cultures mainly concerned with survival).

However, there were evident dangers in the programmed society, mainly stemming from an increasingly perceived need for systems to regulate and control individual activities (a not entirely trivial example was the proliferation of traffic control signs on our road networks) and this was giving rise to a special form of popular anguish and resistance. There was a growing gulf separating "the leaders" and "the led"; a growing concentration of power in mass-communication (via the mass-media); and growing apprehension that information and communication—the keys to future well-being—could if wrongly used become the keys to disaster. Furthermore, the trend was virtually identical whether in centrally-planned (socialist) or market-economy (capitalist) societies: the fact that the former were "closed" in comparison with the latter depended on political not social reasons.

The result was that a new ideological (and often "ecological") elite was today becoming ever more prominent, tending to displace the bureaucratic administrative and business elite of former years. Programmed society could raise many fears, and would be more and more resisted unless the "programmers" were (in the terms of DISCOVERIES) devoted to the humane use of human ideas. In the field of communication, this meant its use to meet social needs rather than as an instrument to confirm and perpetuate the dominance of the already dominant.

Professor Dumont, presenting as examples two case histories of "acculturation" (the absorption of one culture into another) showed the extent to which the process could lead to ideas and ideals being re-modelled and re-interpreted within their new social contexts. Although there was some danger that acculturation could result in an already dominant culture becoming even more firmly established, the ideal of *mutual acculturation* (equal intercommunication and interpenetration) clearly offered sufficient promise to encourage its pursuit.*

These introductory papers were followed by two parallel "round table" discussions, chaired respectively by Yves Jaigu, Director of the "France Culture" radio programme, and by André Danzin, Director (*inter alia*) of the French *Institut de Recherche d'Informatique et d'Automatique* (IRIA).

The Jaigu panel was mainly concerned with the achievement (or more often the failure) of communication via "the media", and it was immediately evident that the failure rate increased very rapidly with the complexity of the information to be communicated. It was also evident that the majority of "media clients" (television viewers, radio listeners, newspaper readers) were usually far more interested in being entertained than in receiving information or instruction.

*This may suggest that the eventual evolution of a single "world culture", based on universally accepted beliefs, values and rules, could be regarded as the ultimate sociological utopia, despite the regimentation of progress that would probably result.

This inevitably raised the question of whether the media should be subservient to majority tastes, catering only for popular demands, or whether they should endeavour to instruct and so improve the general intellectual level. The dilemma was very real for those providing a public service (often at public expense) and also embraced questions of how far the media should go in attempting to present complex matters in simplified form.

The discussion provoked a particularly interesting comment from Professor Boris Rybak of the Paris *Institut d'Etudes Linguistiques et Phonétique*, and Director of "NATO-ASI sur la Technologie", who spoke of the considerable dangers of assuming that majority opinions, because they were of the majority, were necessarily the most deserving of attention. In fact, the exponential advance of knowledge ("information explosion") had led to ever widening gaps between those who understood, those who misunderstood, and those who were "simply deaf"; while attempts to remedy the situation—which must involve considerable effort by those willing to become better informed—tended to be either disregarded by the majority, or to be shunned as part of an indoctrination campaign to confirm power and authority on those already in their possession. This necessarily meant that people who were both responsible and knowledgeable were usually in the minority, although their opinions could well be more valid than those which overrode them. Proliferation of and in the mass-media did nothing to ease the situation, and Professor Rybak went so far as to assert that our survival could depend on our timely diversion from reliance on majority opinions.

This and other interventions (in particular from Professor Ulf Himmelstrand, of Uppsala University, Sweden, who pointed out that all communication, when received was interpreted by reference to any existing framework of connected information*) led M. Jaigu to pose the fundamental question whether the mass of the people should be "protected from the dangers of over-

*This was why an uninformed public was unable to assess satisfactorily an item of information in a novel or unexplored field . . . or in a field where there had been previous misinformation.

information'', on the grounds that they were incapable of correct comprehension (whatever that might be), or whether a policy of maximum mass-education and total information (a ''transparent society'') was to be preferred. The former course must inevitably lead to an elitist-run society, which although capable of great efficiency might well be questioned as a true fulfilment of man's destiny. On the other hand there were clear risks in seeking full transparency, firstly because its achievement could never be guaranteed, and secondly because of the immense pitfalls of the byways should the main road being followed be lost or prove too difficult.

The panel discussion chaired by André Danzin was concerned with a specific aspect of ''programmed society'', namely the many and diverse applications of computers and computer-based information systems. As a preliminary to the debate, a lawyer (M. L. Mehl, Maître de Requêtes in the French Conseil d'Etat), a medical doctor (Professor J. P. Lévy, Cochin Hospital, Paris), and a pure mathematician (Professor J. L. Lions, Collège de France) each gave a short appreciation of how he saw the impact of computers in his profession. Professor Lions who, as Secretary General of the International Mathematical Union, was evidently presenting views widely held by his colleagues, made it clear that they were hardly starry-eyed about computers or computer science. Nevertheless, all mathematicians recognised that the computer (due to its very high speed of operation) had invaluable applications in calculational methods that had long been proposed but, because they would have been too tedious (and expensive) in ''manual'' use, had not previously been adopted.

It was also generally recognised that computers offered some very important advantages in modelling complex social and similar systems, but this did not necessarily mean they were possessed of super-mental powers, or merited such descriptions as artificial brains. From the purely mathematical viewpoint, real intelligence was still very far from the artificial variety.

Next, Professor Lévy gave some details of a computer-based medical data-bank and information-processing system, which was being developed in France and was already proving useful in providing rapid and accurate diagnoses from predetermined standard medical tests. However, the resulting substitution of totally iterative electro-mechanical processes for the long-established "art and magic" of the doctor (let alone the relationships between doctor and his patients) had hardly been welcomed by the medical profession which was neither enthusiastic nor optimistic towards expanded use of computer-based diagnosis. Furthermore, while such a system could enable the fullest use to be made of existing knowledge, it tended to stifle research into new concepts and aspects of medicine, particularly those which might upset the bureaucracy of a computer-organised medical establishment.

M. Mehl, as President of the French *Centre d'Etudes et de Documentation d'Informatique Juridique*, had no difficulty in showing that computer-based information systems could be invaluable in the legal field, where so much depended on precedent and case law and where the information explosion in recent years had been no less violent than in other fields. Not only this, however, but the compilation of rules and principles, applications and judgments from countries where legal customs were traditionally different, could be particularly helpful towards the elimination of practical divergences and the eventual evolution of full uniformity.

M. Mehl also put forward the suggestion that, given suitable modelling of societal "machinery", it should be possible to test by computer the consequences of new draft legislation before it was adopted as law. Although success here would evidently depend on both the representativity of the societal model used and the method of pre-testing adopted, such an application seemed to be generally accepted as possible and probably valuable.

A more contentious suggestion from M. Mehl was that, because computer-maintained records could be made to provide virtually total "transparency" in administration, this could ensure that the treatment of every individual citizen was not only scrupulously fair and impartial, but was publicly seen to be so. Taking as an example the assessment and payment of taxes, he believed that society as a whole, able to examine the detailed application of "the rules" throughout the community and thus to verify that this was equitable and impartial (and above all that it virtually eliminated the possibility of illicit tax avoidance) would welcome the system.

This somewhat utopian concept did not however stimulate, in an imperfect but real world, the universal acclaim that might have been expected in Utopia itself. In a spirited intervention from the floor Professor Harry Gunning (President and Vice-Chancellor of Alberta University, Canada) declared that in his country the advent of computerised personal files, far from being welcomed, had been attacked by many individuals who felt themselves assaulted and victimised by computers, which they sensed were depriving them of both privacy and individuality. The desire to be a unique human being, said Dr. Gunning, was no less strong than the desire to live, and he implied that much of today's resistance for even the most ''innocent'' use of computers was due to this desire being unrecognised.

M. Danzin, with support from M. Mehl, remarked that the much-feared encroachment by computers on personal liberty was somewhat illusory, and further that acceptance of some ''restraints to life'' at this level could be compensated by greater freedom at higher levels. Thus, for example, the very considerable limitations to complete freedom in the use of a road system (speed limits, obligatory directions, etc.) were essential to guarantee to all road users that they could in fact use the roads. The more general theme, that complete freedom for one must mean strict controls for all others, though not evoked was clearly not far below the surface of thought.

On a further suggestion, that computers were essential to resolve the complexity of life today, Professor Murray Eden (National Institutes of Health, Bethesda, USA) commented that computers could very easily make complexity worse. Furthermore, every complex problem which might be resolved could usually be guaranteed to point to an even more complex problem as yet unresolved. There was, in many areas of planning and analysis which could be submitted to computers, a *saturation point* beyond which additional data processing could not be expected to yield improved practical results.

But whatever the reactions to the actual or expected impact of computers on life in all its aspects, there was no doubt that all present at this "Round Table" agreed that tomorrow's society was going to be far more dependent on information and on communication than was today's while on the contrary it must be expected that material possessions and comforts (such as washing machines and motor cars) would be considered less important. Whether representatives of developing countries, still obliged to do their washing in the cold waters of a nearby river and still dependent on animal-drawn carts for their transport, would have agreed with this was not to be determined, for none took part in the session.*

One further point arising from this Round Table should be mentioned. It was made by M. Danzin in his summary presentation the following morning, and was to some extent provoked by Professor Touraine's paper. Referring to the widespread feeling that a computerised society must necessarily also be a centralised society, with ever-increasing concentration of power in the hands of an elite few, M. Danzin pointed out that virtually instantaneous electronic information exchange could equally well—and perhaps more effectively—lead to *devolution* of real (decision-making) authority among a multitude of quite small though interdependent communities. This must mean that the individual, rather than being dominated by the (national or world) state, could expect to preserve both a degree of freedom and a degree of influence in relation to the evolution of society about him. In other words the computer and its associated complex of automation even in "informatique" could and should help mankind towards a new freedom of the spirit as well as material prosperity. If this were not achieved it would not be the fault of the machines.

Returning to more general sociological aspects of evolving society, and perhaps stimulated by Professor Rybak's remarks on

*"Third World" representation in DISCOVERIES Symposia to date, in fact, though certainly not absent has been somewhat slim, despite the probability (United Nations estimates) that over 80% of the total world population by the mid-2100s will be in Third World countries.

the fallibility of majority opinions, Professor Dumont had some comments on other "popular assumptions"—often resulting more from inbred intuition than from scientific or logical reasoning—which inevitably affected basic attitudes to what we called civilisation.

Currently, there was a particular ideological concept which could seriously hamper correspondence between such attitudes and practical reality. This was the concept of a virtually universal egalitarianism. All factors were assumed equal to all other factors; while for the anthropologist, the ethnologist, called on to examine a society or civilisation, the opposing concept of *hierarchisation* was inescapable. "You constantly need to know what is predominant and what is less important", said Professor Dumont, "what facts or effects are primordial, what are secondary".

This "unthinkable" idea of hierarchies and non-equalities was also vital to the study of information, communication and comprehension, and however much our intuition might be opposed to it, it represented a reality which was both inevitable and potentially dangerous. "It is", said the Professor, "precisely because we have in modern society a juxtaposition of deeply-rooted individualist values as well as an effort in search of the holism of a global society . . . an effort to re-invest such society with primordial value, that we may encounter totalitarianism. That, after all, was the origin of Nazism".

He might have added that it could easily be the origin of other forms of intolerant society. Further, that attempted "compulsory egalitarianism" was a sure way to promote hierarchisation in its most destructive form.

Here, perhaps, is the most fundamental dilemma of all fundamental dilemmas for the future of mankind.

The Development of Communication

The first of the two "scene-setting" papers of the second working session of the Paris Symposium was given by Professor Murray Eden, of the US Department of Health, Education and Welfare, under the title "Consequences of Technology for

Communication''. These consequences had in the past related mainly to information *transmission*, being mostly improvements in reproduction of the written word, or more recently in transmission of the spoken word by telegraph, by radio, and now—accompanied by visual images—by television.

But today's technology was more concerned with the *transformation*, presentation and use of information after its receipt, and here the rapidly expanding applications of electronic systems—particularly computer-based systems—was probably more profoundly significant than had been the expansion in the past of printing. These developments, said Professor Eden, had drawn attention to the complex relationships between the needs, priorities and goals of an information sender and of the receiver of that information—which might be similar, complementary or even antagonistic. ''Intention on the part of the participants in communication'' he said ''plays a critical role in shaping both the directions which the technology has taken and the evolution of their social impacts''.

As examples of this the Professor described two recent developments in the United States. The first concerned the phenomenon known as CB (Citizens' Band) Radio, which had manifest itself through the installation of nearly *14 million* low-power VHF (27 Mc/s) transceivers mostly in private motor cars. The ''big business'' aspect of this aside, the really intriguing part-consequence/part-cause was the widespread use of these devices for citizens to establish anonymous radio relationships—talking together of matters which they would never mention within their own circle of recognised friends and relations. Here, surely, was proof that man today needed a confessor no less than in the past.

The second example concerned the introduction of a ''Universal Products Code'' for the labelling of supermarket sales items in such a way that electronic scanning at the check-out desk could provide not only the customer's account but also much additional information notably for continuous ''on-line'' stock control, ordering of new supplies, etc. Here was a case of the fullest possible practical exploitation of initial information by an advanced technology which, although hardly ever understood in detail by the user, nevertheless had his confidence and trust. An interesting observation was that this essential confidence and trust had not been dependent on the technical features or a technical evaluation of the system adopted, but rather on the reputation of a particularly well-known and well-trusted corporation in the data-processing field (IBM) which had been responsible for the basic features of the system.

The second main paper of this session was given by Professor R. Narasimhan, of the *Tata Institute of Fundamental Research* in Bombay (India), who presented a detailed account of computer-modelling techniques applied to the study of human behaviour dependent on language*. As noted by Pavlov, animals relate to their environment directly, via a complex "sensory-motor" system, while human beings make use of a "secondary signalling system" namely language which profoundly affects apprehension and comprehension of a given situation, enabling—*inter alia*—experience to be stored for future reference. Language is acquired by example, by identifying sound patterns with certain situations and not, as traditionally supposed, through "internalised" grammatical systems. This is of fundamental importance when simulating situations from which to deduce behavioural patterns: if a computer is fed with incorrect assumptions or data as basic information, it cannot normally correct the error and so must deliver false conclusions.

Professor Narasimhan emphasised a matter which was not perhaps recognised by everyone in the Symposium, and which might even have been denied by some: namely that the electronic digital computer was "basically stupid", and derived any "intelligence" it might appear to possess from its creator, man. This was not to reject the concept of Artificial Intelligence, though it was a valuable reminder that—at least among some thinkers—the famous dictum of Descartes remained valid.

The first of the two following panel discussions, under the title "Communication between Human Beings" had much relevance to Professor Narasimhan's paper. It was enlivened at the start by

*Paper title: Communication and understanding: a comparative approach to language behaviour.

the projection of a teletape recording showing the reactions of a 4-day old baby to sensory (mother's voice) and tactile (mother's touch) communication. This preceded a short paper by Professor T. Ishii of Tokyo University, which dealt with experiments on pattern recognition in living beings and attempts to model these on computers. Significant results had shown that babies were able to appreciate their mother's voice long before birth*—also that communication was "two-way", in that her baby's cry will stimulate a mother's secretion of milk. Understanding the mechanisms of these (clearly not unreasonable) phenomena, and those governing the baby's subsequent acquisition of its first fragments of language and language structure, are of great interest for the developers of computer simulation of language learning, right to the most sophisticated levels of (returning to Professor Narasimhan's paper) what is becoming known as Artificial Intelligence.

The genetic transmission of (genetic) information has now been examined and investigated by very many researchers, prominent among them being Professor Marcello Siniscalco of the *Sloan Kettering Institute for Cancer Research* in New York, who gave a brief discourse on communication within molecular biology (storage, replication, translation and transmission of genetic information) and its relationship with the ecological environment in which it must take place. Professor Siniscalco brought a breath of "real" (i.e. non-modelled) science to the discussion, not to mention the welcome observation that "cultural communication" can be looked upon as the most sophisticated biological feature developed by *Homo sapiens* (over and above societal instincts equivalent to those in other mammals) and that this feature is seen by many as the factor which will eventually enable mankind to direct the course of his own evolution.

Professor S. Aida of the Tokyo University of Electro-Communications (and incidentally Secretary General of the Paris Symposium) noted the distinction between "internal" human communications (physically inside the body) and communication

*Some have maintained that this is true from the moment of conception.

between the body and the environment. Assembling a valid model of the combined communication structure was a fascinating, important and difficult task, additionally complicated by the fact that traditional regional customs (he cited the American practice of telephoning daily between husband and wife to confirm continuing affection) could also affect the overall structure.

Next Professor Eduardo Caianiello (Salerno University) reminded his colleagues that—as had been said many times already—communication was above all a matter of *translation* of information, and that without this there was in fact no communication. He was today, he said, after some 40 years of scientific activity, trying to "unlearn" much of the conditioning which his early training had given him, for he realised how much this conditioning could affect the way in which he would *translate* incoming information. As a simple (though certainly not trivial) illustration of the different interpretations which could be put on the same event, Professor Caianiello remarked how a geologist, once lost in the snows of Canada, had said to his Indian guides that he believed they were all lost. "Not so", came the reply, "we know very well that we are here: it is our tent that is lost".

Finally in this discussion, Professor Barrington Nevitt of Toronto University (well-remembered from the previous DISCOVERIES Symposium in Rome) took further forward the appreciation by babies before birth of their mother's voice, saying that a baby's first knowledge and mental *formulation* of his mother tongue (so correctly termed!) had also been shown to begin before birth.*

Professor Nevitt saw the human learning process as comprising three simple factors: rhythm, rhyme, and reason—which were acquired in the order given. He also insisted—as he had done at previous meetings—on the importance of distinguishing between *concepts* and *percepts*, and of recognising how much the former depended on the framework (cultural, intellectual, mathematical, etc.) in which they were conceived. All "learning models" were

*This would seem to restrict those validly claiming to have several mother tongues to children of linguists.

necessarily conceptual also, and he pleaded for more attention, in trying to understand communication between living beings, to *percepts*—not only visual or audibly acquired percepts, but also those involving the important sense of touch: "We must", Professor Nevitt concluded, "understand what the communication process is doing to us, or we can never hope to control it".

The next panel discussion was effectively an extension of the previous session's "round table" on relationships between the media (the instruments of communication) and information communicated. In that earlier discussion it was noted that most people looked to the media primarily for entertainment, and Percy Tannenbaum (BBC London) not only believed this was a correct order of priority, but that in fact it was the order normally observed.

Without contradicting this Professor Kurt Lang (Oxford University) insisted on the importance of the media for political information and the formation of public opinion. Information must inevitably play a considerable part in establishing, modifying or qualifying our opinions, political or otherwise, and it is via the media (and usually the mass media) that we normally acquire this information.

Dr. A. Willener (of the *Institute for the Sociology of Mass Communication* in Lausanne, Switzerland) described various practical experiments with local television, which evidently must usually have greater relevance for the local population than could national networks. Experience to date in several countries (on both sides of the Atlantic) had shown that local TV was by no means as simple as it appeared at first sight: it was often difficult (and costly) to cater specially for local tastes; in industrial areas most viewers were most of the day at work and therefore not available either to produce programme material or even to watch the programmes; and both stage and transmitting equipment was expensive.

Professor Reikichi Shirane, of the Telecommunications Science Foundation in Tokyo, spoke of increasing interest in *intercommunication* between media and the people (i.e. feedback) to enable the wishes of particular sectors of the population to be made known and then accommodated. He recognised that this was not normally possible for the true mass-media, but it was being adopted—with some success—by the "medium-media".

Finally, M. Mario Borillo, Research Director on "Informatique" of the French CNRS in Marseille, postulated some wider uses for computers in Mass Communication, and also in scientific communications over wide networks. Bringing the discussion to a close in this way was a powerful, though perhaps unintentional, reminder of the fascination that "computerology" seems to have for so many people in so many contexts. It is of course undeniable that these machines can be—and are being already—of immense help in very many areas and undertakings, and it would have been impossible for DISCOVERIES to examine the subject of *Communication* without much attention being given to them. Nevertheless, neither the papers nor the various discussions in this second session of the Symposium offered any real evidence that communication in human society is other than a fundamentally (some might say by definition) human affair.*

Communication, Understanding and Society

Man's curiosity about himself surely dates from his first arrival on earth, equally surely extending yet further into the past in a simultaneous search for an explanation of his origin and for his *raison d'être.* The third working session of the Paris Symposium, looking at the most advanced attempts to analyse and simulate human communication, no less surely marked man's continuing dogged persistence—and growing success—in understanding his own inbuilt secrets and constructing analogous competitors—which (at least in speed and accuracy) have already shown

*Although it is commonplace in modern technology to speak of "communication between machines", the communication in fact achieves human significance (significance for "human activity") only when it is finally interpreted in a human mind and expressed as consequent human action. Furthermore, until there has been such action, there is no tangible evidence that communication has been achieved.

their superiority at some levels of logical evaluation and "thought".

The session began with two profound intellectual exercises in the challenging tradition of their *alma mater*, the Sorbonne University (now known as various numbered units of the University of Paris). In two largely complementary papers, Professor J. F. Le Ny and Professor J. C. Simon provided a high-level introduction to the theory and practice of Communication Mechanics.

Professor Le Ny's paper was in fact a treatise on comprehension—an activity fundamental to communication as had already been recognised in the previous sessions*. Assuming that human comprehension is (as seemingly it must be) an "algorithmic activity", Professor Le Ny reviewed recent work by a number of specialist teams including his own which essentially suggested:

(1) that basic comprehension algorithms are stored in the human (and indeed all other animal) brain;
(2) that the operation (execution) of the algorithms depends on a form of "semantic memory"; and therefore
(3) that the ability to comprehend a "message" in the way intended by the person transmitting it must depend on equivalence between the receiver's and the transmitter's semantic memory.

Applying these ideas to class and group communication, the Professor indicated the importance for communication of social behaviour (a form of communication in itself), and the importance of certain innate or intuitive abilities as opposed to those which might be acquired through (e.g.) learning.

Professor Simon carried this basic analysis forward to present a practical demonstration—through diagrams, projections, and a remarkable recording of a totally artificial voice generated in a mini-computer from a typed input message. This most impressive performance, although it seemed still a very long way from "true

*Though not, unfortunately, always recognised elsewhere.

artificial intelligence''* was an invaluable display, for those not frequently or closely in touch with research of this nature, of the tremendous advances which have been made not only in computer programming but also in the computing machines themselves. In particular, the ability to carry out extremely complex (and/or tedious) computations in ''on-line'' central equipment for, e.g., exploratory satellites investigating the solar system or further afield, is a dazzling tribute to the non-artificial intelligence of Man himself.

It should nevertheless perhaps be noted (as remarked by the mathematician Professor Lions in earlier discussion) that the tremendous advantages offered by modern computers—and by the now extremely sophisticated programs for their use as well as ''peripheral'' equipment which a few years ago would have been unrealisable even if conceivable†—still appear to be no more than very advanced ''mechanical'' copies of certain human brain functions. The idea that machines can be endowed with the equivalent of mental *initiative*—or even decision-making powers other than those built in by their human designers, still seemed to most of those present (if not all) difficult to justify either through experience or through logic.

Descending from such heights of cerebral exercise to more earthly intellectual and cultural problems, at least some Symposium participants must have been struck by the contrast between the microscopic analysis of voice patterns displayed on an oscillograph and the macroscopic approximativity of ''ordinary'' conversation in the ''round table'' discussions which followed. These dealt with ''Problems of Group Acceptance of Information'', and ''Cultural Problems of Technology Transfer''.

The ''Group Acceptance'' discussion was opened by Professor

*However that might be defined.

†It is for reflection (though this was probably outside the scope of the Symposium) that the development of technical equipment and of concepts for its use and further development, generally go hand-in-hand. Modern large-capacity digital computers would surely have been inconceivable without the transistor, whose arrival immediately opened fields of logical simulation which themselves had previously been inconceivable.

Henry Tajfel of Bristol University (UK), who attacked his subject from two viewpoints:

(1) acceptance of information as a function of the (accepting) group's internal composition, structure, function, goals and ecology; and
(2) acceptance as a function of the group's relations with other social groups within a wider social structure.

He believed the second issue to be of much the greater importance, and was particularly concerned by the possibility of complete communication blockage when conflicts occurred between groups which were basically interdependent. Unfortunately, the causes of conflict frequently included a strong desire by such interdependent groups to represent themselves as independent, an objective clearly irreconcilable with reality.

Other causes of conflict cited by other speakers included those based on the difficulty of admitting "the truth" about some particular development or event which seemed culturally unacceptable. For example Professor Pietro Omodeo (Padua University) described the rejection, some centuries ago, by doctors and pharmacists of a new medical discovery which ran contrary to the medical beliefs of the day*. Somewhat related was the modern difficulty—perhaps encouraged by many peoples' concept of the mass-media as a type of propaganda machine—of convincing the population of, for example, the real magnitudes of comparative risks associated with different methods for producing electrical power. As Professor Torgny Segerstedt (Uppsala University) remarked: "When we receive a message which is relevant to an old custom we generally ask a person we regard as an authority, but we do not seem to have any 'real' authorities with regard to new fields of information".

A particularly interesting contribution was that of Professor Tsujimura, one of the few participants in the Symposium to examine differences between Western and Eastern thought—the

*Shades of Galileo!

bringing together of which is among the objectives of DISCOVERIES. Professor Tsujimura described the influence of Zen on the Japanese way of life and attitudes, making it clear that these attitudes could be of particular assistance when contemplating revolutionary and perhaps unexpected changes in the material world. The West follows Descartes' "I think therefore I am" in recognising the Ego as a fact of life through which all perceptions are made. Zen, on the other hand, does away with the notion of Ego, this being most obvious in Japanese Zen art. Objects are here represented in such a way as to capture their "essence", whilst the personality of the artist is totally bypassed.

An immediate and strong impression from the discussion of cultural problems in technology transfers was that such transfers often (usually?) concerned technology that the advanced countries believed the developing countries *ought to have* (rather than what the developing countries were themselves seeking), the transfers being frequently favoured because of their indirect commercial profitability for the manufacturing and other organisations involved. Moreover, on occasions when priority was given to evident needs (one example cited was the supply of radio communications equipment to medical posts in Nicaragua, to enable clinical advice to be obtained rapidly from a central control station) there was in fact very little real technology transferred, but rather a series of operating "black boxes" requiring minimal expertise for their use. In these circumstances it was hardly surprising that Professor Donald Ekong (University of Port Harcourt, Nigeria) should call for greater emphasis on the transfer of basic scientific knowledge and competence rather than "packaged technology".

Yet such "basic transfers" could also raise severe problems, as was emphasised by Professor Zvonimir Damjanovic (Director of the Multidisciplinary Studies Centre in Belgrade, Yugoslavia). In cases where educational levels—cultural as well as technological—were widely different there could be active opposition from those intended to benefit from the transfer, in other words a form of "group rejection".

Here Professor Damjanovic saw a close relationship with the questions discussed by Professor Tajfel, who had himself been more concerned with the problems of group rejection of information than with those of acceptance*.

Professor Tajfel also pointed out (as was evident from Professor Le Ny's paper) that there could be important differences between the mechanisms—and consequences—of individual and group acceptance or rejection of information—and he cited examples making it clear that group behaviour was (naturally) dependent on the nature of the group and that (unfortunately) it was by no means always rational, scientifically or logically.

A most instructive contribution to the discussion of acceptance or rejection of information came from Professor Harry E. Gunning (President of Alberta University, Canada) who reported on some practical experiments which suggested that the human brain will normally and automatically (sub-consciously) prepare for the rejection of *any* information communicated to it because (in the Professor's words) ". . . attempting to convey information is an act of ego-invasion and, as such, if we attempt to invade the ego of another there is a natural tendency to resist".† This must mean that the acceptance of information normally involves a conscious effort, doubtless accompanied by a tendency to select and perhaps modify what is accepted. A further conclusion must be that attempts to *force* information on to others are virtually certain to fail, and that the most effective communication is likely to result from "working together and *sharing* information", so taking advantage of the backgrounds

*At least some participants, in fact, maintained that the only problem of group acceptance of information as such (i.e. excluding problems which might result from acceptance of "unhelpful" information) occurred when the acceptance did not take place.

†The practical evidence cited by Professor Gunning was the discovery that, when a person A attempts to talk to a person B, the vocal chords of person B immediately begin vibrating in subsonic patterns apparently opposed to the patterns stimulated by person A.

It is to be noted that such a reaction could hardly be possible in a practitioner of Zen, as described by Professor Tsujimura, if this Eastern discipline does in fact eliminate the concept of ego. Could it be that the practice of Zen results in substantial modification of the operational structure and processes of the brain?

(concepts, semantic memories) of both, or all, the persons involved.

In this connection Professor Simon was able to report on some exciting attempts to make computer models which could receive information and *learn*. This was proving extraordinarily difficult: in particular, unless ''incoming'' information could be subjected to an important degree of selection and ''filtration'', instead of *learning* the model fell into a state equivalent to either epilepsy or catalepsy. This suggested that human beings, who suffered such effects comparatively rarely, were equipped with remarkable protection against incoming information—circumstantial confirmation, perhaps, of Professor Gunning's ''real life'' observations, and further suggesting that rejection of a great deal of information, whether by individual or by groups, was probably essential for the maintenance of social stability.

Endeavouring to summarise the causes of group non-acceptance of information, Professor Tajfel concluded that four main factors were involved:

(1) clash of interests (i.e. clash with the *interests of the group*);
(2) clash of values (i.e. clash with the *values of the group*);
(3) conflicts between values and structural ''predicaments'' (situations), where the predicaments were over-riding and could not be altered to resolve the conflicts;
(4) clash with ''new ignorance'' (inability to understand evolving developments—particularly those of advanced technology).

Unfortunately, the identification of these factors was by no means a guarantee that the conflicts involved were resolvable; however the Professor believed such resolution might be achieved by a combination of *assimilation* (the modification of new information in already existing cognitive structures according to the nature of these structures) and *accommodation* (the modification of the cognitive structures according to the nature of the new information received). Both these processes were large-scale and in competition, and it was necessary to establish a balance between them if the new information was not to be automatically rejected—or lost. For if assimilation predominated over accommodation, the information was likely to be distorted beyond recognition, while if accommodation were predominant, then the previously existing structures would be destroyed and would disappear.

Communication and Socio-economic Phenomena

Opening the fourth working session of the Paris Symposium, Professor Ithiel de Sola Pool (Massachusetts Institute of Technology) recalled Professor Simon's earlier remark that nearly 50% of all working occupations were now concerned with the production, treatment and dissemination of information.

There were four main reasons for this evolution:

(1) growing economic interdependence, leading to a "global economy";
(2) rapid and efficient (and cheap) transport of persons and things;
(3) electronic communication; and
(4) "intelligent" machines.

Of these the most immediately significant were electronic communication especially via geostationary satellites, and transport—even though the former was likely eventually to make "business" transport less and less necessary. One result would be the inevitable collapse of national frontier controls on information transmission, and on services dependent on this such as news diffusion, interchange of cultural, technical and economic data and, for example, banking (banks anywhere in the world would be accessible "to anyone who can pick up a telephone anywhere").

Professor Jean Voge (French Direction Générale de Télécommunications, Paris) added quantitative data to Professor de Sola Pool's information, in particular developing rules for "optimum" structures in communications networks and showing that a system of exchanges at several levels (subscribers being connected to "primary" exchanges which were themselves interconnected through "high-order group exchanges" at one or more levels) required less commutable connections and consumed less power than if a single large centralised exchange were used. This demonstration that a degree of decentralisation led to greater efficiency clearly had relevance in other fields, notably that of agglomerations of "population units" in large cities as opposed to their distribution more evenly over much wider areas. The compelling reason of the past for concentrating population in cities (the need for frequent and rapid personal contacts with many other citizens) was disappearing as electronic communication made personal contacts a matter of choice rather than of necessity.

Concerning information content in production, Dr. Voge cited figures, mainly from the United States, indicating that an optimum content (information "quota" or "ratio") was 50%, and that in the US this had now been very nearly achieved. The more advanced countries elsewhere in the world were in pursuit.

Both of these papers pointed to the important benefits—material and "moral"—to be expected from replacing today's concentration in cities by the gradual decentralisation of society which modern communications have made possible. One somewhat paradoxical observation, however, was also to be noted, namely that increasing unification of communications structures (e.g. through satellites and even through large though "distributed" computer systems) could not only lead to greater physical decentralisation of much of our currently centralised society; it could equally lead to even *further* centralisation, which at least initially would probably exhibit increased efficiency.

These presentations led naturally to a discussion opened by Professor Ilya Prigogine (Free University of Brussels), who gave a brief exposition of his celebrated "town" model of a complex societal unit, interacting with its environment and at the same time pursuing an internal evolution due to such developments as expanding population or increasing road traffic. Towns in this sense are "dissipative structures", and when some part of the system is disturbed by a "communication", resulting non-linearities in the various inter-related evolutionary paths of the system may provoke unforeseen changes which, in Professor Prigogine's terms, are "bifurcations". Without such non-linearities the system would not of course—and could not—remain a "living system".*

This type of analytical approach was mentioned in a case study of the Belgian capital Brussels, presented by Professor P. Laconte (Louvain Catholic University), who described the very substantial changes resulting from the installation of EEC Headquarters in Brussels in 1960, and the subsequent development of the city as an international centre of administration and business. Professor Laconte's account of the present situation of Brussels clearly represented the many and complex problems associated with such a development and amply demonstrated the need for and value of dynamic analyses of this type. It also offered practical confirmation of some of the conclusions of Professors de Sola Pool and Voge concerning the relative merits of concentration and decentralisation in heavily-populated areas.

In a characteristically philosophical comment on these questions, Professor Eduardo Caianiello offered a timely reminder that, whatever the conclusions of scientific studies and analyses—which should certainly be used as guides towards sound future planning—they should not be allowed to exclude considerations of human wishes and human happiness—which meant accepting that *optimum* technological efficiency would not always or necessarily coincide with the *maximum* such efficiency possible.

While not contesting an earlier remark by Professor de Sola Pool, that man's survival depended on the use he made of technology while his happiness depended on the use he made of survival, Professor Caianiello believed that survival without happiness would be as meaningless as happiness without survival would be impossible.

*Those wishing to pursue Professor Prigogine's theories in greater detail are recommended to his paper "Metamorphoses de la Science: Culture et Science aujourd'hui" (in French) prepared for the previous DISCOVERIES Symposium in Rome.

Professor Umberto Pellegrini, President of the Italian Federation of Scientific and Technical Associations, commented on the major changes in social and economic structures resulting from the evolution of pre-industrial society into industrial society, and from the latter to post-industrial society. In a somewhat similar analysis to those of Professors de Sola Pool and Voge, Professor Pellegrini saw three main levels of communications-related activity in post-industrial society, namely:

(1) *informational activity*—information being a basic resource which in post-industrial society would play a similar role to that of energy in industrial society;
(2) *"intellectual technology"*—based on information, abstract theory and science, these being substituted for the intuition and empiricism characteristic of industrial society; and
(3) *a new "working class"* of professionals and technicians trained in the applied sciences.

These developments would certainly, as Professor de Sola Pool had indicated, radically reduce the significance of national boundaries, and would lead to new types of transnational social aggregations and interactions. Such new relationships, dependent on electronic computers and telecommunication systems ("téléinformatique") would allow the operation of many decentralised but highly productive units, every one with individual access to the information resources of the entire world.

The delicate question of industrial relations in the foreseen "programmed society" (a society which is certainly not yet universally accepted as representing the humane use of human ideas) was addressed by Dr. Fred Margulies, of the Austrian Federation of Trade Unions. Immediately concerned with measures to preserve cohesion, cooperation and satisfaction in relations between Professor Pellegrini's "new working class" and society as a whole, Dr. Margulies suggested that these measures should include:

(1) improved and speedy communications to and among the

workforce representatives of details and probable implications of foreseen economic and technological changes;

(2) participation of workforce representatives in decision-making affecting the future of all;

(3) recognition by workers and their unions, as well as by managers, that every technological and organisational problem offers alternatives, that these frequently (usually?) involve conflicting interests, and therefore that the question of who decides what should be done is no less important than what is decided;

(4) recognition by managers that the greater satisfaction among their workers which would result from more opportunities to take decisions and make their own choices would almost certainly lead to better work while—on the other hand—sophisticated automated systems were unlikely to work satisfactorily if controlled by dissatisfied workers;

(5) recognition by society that human values cannot be reduced to simple production figures and financial returns: that the most important of all human values, in fact, is human dignity.*

No-one who heard these suggestions was prepared to dispute them, nor that they concerned, beyond the humane use of human ideas, the humane use of human beings themselves. Nevertheless, and by no means for the first time in presentations of this nature, it was not easy to suppress an underlying feeling that "real life" was less simple and straightforward than here suggested. An example, perhaps, of "message refusal" due to conflict with pre-existing understanding?

This fourth working session, possibly because it was the fourth and therefore benefited from the "preparation" of the three preceding ones, was a particularly stimulating exercise towards a combined interdisciplinary approach to the technological, socio-

*cf. Professor Caianiello's remarks on happiness.

logical and *human* problems of a Communications Society. In a further presentation, Professor Prigogine noted how the combined application of many sciences, particularly physics and mathematics, was today enabling the overall nature of complex systems such as human society to be seen with new clarity. The remarkable complexity of these systems was also, for the first time, being seen in detail, as was the impossibility of breaking them down into simpler systems which could be satisfactorily represented by simplistic models.

Their evolution was continuous—yet continuous only in the sense that it comprised a series of sudden, non-linear changes—or in the Professor's terminology, "bifurcations". All progress in fact must depend on these non-linearities—without them the future would be entirely settled and determined by the past—evolution would become a banality.

Professor Caianiello, while not disagreeing with the general direction of these enunciations (i.e. that mankind, with the help of science, was constantly advancing towards a better understanding of his environment, including his own society) clearly believed there was still a very long way to go. "We do not", he said, "really know enough, if we know at all, of the laws that make human societies of various degrees, qualities and so on; nor do we know about many essential facts that still concern the perception by the individual of some message that may come to him".

The Professor was also unprepared to dismiss simplistic models as without value in studying very complex systems. While recognising that they could be very crude he believed that this could often be an advantage: indeed, much past progress in investigating the complex processes of nature would have been more difficult or even impossible without the use of crude and simplistic models.*

However much more sophisticated models—such as those of Professor Prigogine's "nourished equilibria"† (to which

*The (momentary) assumption that certain variable parameters in a relationship were invariable was a classic way of arriving at crude models which, despite their crudeness, could be of remarkable help in understanding aspects of very complex processes and systems.

†For details see reference paper cited earlier.

Professor Caianiello paid particular tribute) were also more and more necessary as understanding advanced and, as was clear from the papers and discussions in the present Symposium, many such models would not be realisable without the help of modern computer and information-processing systems.

Surveying some of the other remarks and presentations during the session, the Professor discussed the impact of so-called post-industrial society on the labour force which that society had inherited from the preceding industrial society. Both Professor Pellegrini and Dr. Margulies had drawn attention to the decreasing demand for "elementary" (manual) labour, and the increasing professionalism in the new "intellectual technology".

Professor Voge had indicated how advanced countries were approaching a situation where 50% of all production would comprise informational activity. Clearly this implied very significant changes for traditional workforces, and there was a developing problem of how, in a world of information and automation, those displaced by machines should occupy their lives. "We have", said Professor Caianiello, "to prepare more students in our highschools, experts in our universities, so that there may be places where these people find the right occupation and a job". He continued: "They will be an essential means for human development provided we are careful enough to take care of such things now", and in this he was echoing the sentiments of all who, seeing the path along which modern society was now set, were apprehensive of the possible consequences for a substantial sector of every community. For despite the encouragements of additional education and more intellectual work, not everyone in that sector would necessarily wish to follow this path, yet they would be unable to avoid its effects and implications. Perhaps we should see a massive return to traditional handicrafts, themselves with a high information content and offering personal satisfaction and fulfilment not to be found on the production line.

Professor de Sola Pool also referred to the complex systems models developed by Professor Prigogine, pointing out that these did not unambiguously favour either centralisation or decentralisation as desirable future policies. Such "simple polarity'' was unrealistic: for example did the growth of modern conglomerate organisations in industry or elsewhere represent more centralisation, or less? ''The only way they can grow bigger'', said the Professor, ''is by adding levels into the hierarchy and allowing a much larger degree of decentralisation of components''. A conclusion clearly in line with those of Professor Voge concerning telephone systems.

The session ended with further discussion of the ''false controversy'' between centralisation and decentralisation. Although at first sight these might appear as alternatives, further examination showed them in fact to be complementary, and the future search should therefore be to find the optimum combination. The general trend of thought seemed to be towards some system of ''dispersed hierarchies'', leaving matters of no more than local concern to be decided locally, but recognising that others (food supply, raw materials, energy, international informatique, or the problems of pollution) should be regulated at what Professor Voge called ''global level''—for (as he said) we can visualise the world as a ''global village, made up of a large number of micro-societies''. In such a system, he believed, most of the power would exist and be exercised at the lower levels, but some would necessarily remain—and even be further concentrated—in the top echelons of the hierarchical pyramid.

Monsieur André Danzin raised one further matter. Recalling the problems expected to develop for many people of how they should occupy their lives when—in the society of the future—the ''mechanics'' of living required only a small part of their daily effort, he believed quite separate structures could be created in which people participated voluntarily (or at least without immediate material reward) in projects designed to develop and improve their environment, their participation in the world about them, in fact their contribution to society.

Here, surely, was a novel concept of communication in the post-industrial world—communication in the form of *service* to the community as a principle of life.

Novel? Or one of the oldest and noblest concepts in history?

Communication Achieved

The closing session of the Paris Symposium comprised a review, in the form of a final panel discussion, of the previous working sessions. Under the presidency of Professor Jean-Claude Simon (the Symposium's Scientific Committee Chairman) it was honoured by the presence of Monsieur M. Turpin of the French Ministry for Industry, who brought a message of congratulation and encouragement from the Minister, Monsieur André Giraud.

In reply to an introductory remark by Professor Simon, Dr. Shimoda, President of the Honda Foundation pronounced the Foundation extremely satisfied with the meeting, and paid tribute to all those who had contributed to its success. He emphasised that the Foundation had not expected "instant answers" or concrete conclusions, but rather guidance for its future work. "You", he said, with the humility typical of the oriental approach, "you will decide what direction should be taken".

It was Professor Prigogine who, in taking this invitation further, echoed the thoughts of most (if not all) present by asking how the sought-after "concrete action" might eventually be distinguished and launched. He believed there was wide agreement that communications and "informatique" must necessarily be involved, for they were essential to any societal structure. Furthermore, in a properly functioning democracy they must be fully and recognisably "open". This was where difficulties could arise, for communication on technical matters with non-technicians raised many formidable problems—whose non-resolution was at least partly the cause of the present "crisis in science".* It was of course vital that those of the intellectual disciplines, as represented in the Paris meeting, should communicate successfully among themselves, but it was not enough: communication with "the public" was no less important—and indeed with governmental authorities and parlia-

*Opposition to technological or indeed any other progress is probably as old as humanity, and those who suffer from it today might occasionally reflect that at least their professional careers and *lives* are rarely threatened. Considering the possible practical consequences of some modern developments, this is surely a triumph for tolerance if not for understanding.

mentarians, who were usually the source of the investments without which scientific work would die. ''A Government which pays big research bills'', said Professor Prigogine, ''has a right to the results''.

Several members of the panel, notably Professor Alfred Sauvy (Collège de France, Paris) and Professor Torgny Segerstedt (Uppsala University, Sweden) reflected on the somewhat esoteric nature of the Paris meeting, drawing attention to the dangers of concentrating thought on very advanced theoretical and philosophical questions while neglecting the practical needs of survival. Professor Segerstedt also stressed the need to keep in communication with (and keep the confidence of) the younger generations of intellectuals, who would soon be shouldering the immense responsibilities of decision-making in fields non-existent in their parents' youth.

Professor Caianiello, who had been Vice-Chairman of the Symposium's Scientific Committee, defended the sophistication of the Paris approach by pointing out that communication was in fact an extremely sophisticated subject, and one which was vital to the future of the DISCOVERIES Project itself. Many other meetings around the world were devoted to ''immediate aspects of futurology'' (if that was not a paradoxical statement) and the Professor paid tribute to the Club of Rome and similar study enterprises. Where he believed DISCOVERIES differed from these was in its proclaimed objective of trying to pool oriental and occidental knowledge, wisdom and philosophy for the benefit of world civilisation as a whole. To a large degree, DISCOVERIES was concerned with the means towards the ends that were sought, and it was to be noted that history had shown the results of undertakings to be very often far removed from those envisaged. DISCOVERIES must try to avoid this, and communication was an essential tool not only for forward progress but, and in particular, for providing the feedback loops necessary for maintaining stability without losing direction.

Dr. Jean Cloutier (International Institute of Communication, Montreal) then performed a brilliant analysis of the Paris Symposium itself—analysing the use of the time available despite extra hours, he found the time for discussion, as opposed to listening to papers, too limited*; the success of communication at the two (quite distinct) levels of intellectual exchange and language exchange (translation); and the intervention, whether intentional or not, of many other factors such as participants' individual intuitions, psychologies, and approaches to "public speaking". In particular, Dr. Cloutier remarked on the distortions of information which inevitably took place as speeches were recorded, translated, retransmitted and published—and he saw these various stages as giving rise to "noise" in which meanings of insufficient amplitudes could be lost. He believed there was need for a profound and extensive study of the "basic mechanics of communication", even those which at first sight seemed elementary.

Professor Barrington Nevitt (Toronto University) agreed with the "noise" analogy but believed that, in addition to the sources mentioned, some noise was due to differences in cultures, notably between those of East and West. He called for greater efforts by representatives of each to understand the other without, as he said "reducing one to the other". Professor Prigogine appeared to be unsure of this, saying that differences between some eastern cultures were at least as great as those between eastern and western: he believed an important contribution of science to intercultural understanding was precisely that the language of science—at least for the present—was more rigorously defined and more universal.†

Professor Alwyn Scott (Wisconsin University, USA) while not disagreeing with this general thesis, seemed unconvinced that the language of science retained its precision unimpaired when it was applied, for example, in psychosociological fields by psychosociologists. Professor Scott further questioned Professor Prigogine concerning the applications of his own terminologies (and

*This is an almost universal failing in symposia gathering more than a few participants. It is particularly irritating when much valuable time is taken by speakers who *read in full* their papers, the texts of which have been (or should have been) issued in time for advance study by all those taking part—so leading to fuller and better-informed discussion.

†In fact, these two professorial opinions must surely be complementary rather than mutually exclusive.

theories) in the realm of pure physics, suggesting that these might in some instances represent an unnecessarily complicated approach. This was not the first time during the Symposium that a participant with basic training in physics or engineering had implicitly championed use of the simplest possible explanation or model of a phenomenon; and those who had been in Rome for the previous Symposium were reminded of a remark by Professor Simon questioning the *necessary* validity of ''Occam's Razor''.* Professor Prigogine, while not denying the value of ''simplistic models'' without which much scientific understanding could not have evolved, believed that today physicists (and others) were recognising that ''reality'' was usually extremely complex, and that much effort could be lost in seeking a simplistic general pattern which did not exist. A further comment (from Professor Marcello Siniscalco, Sloan Kettering Institute, New York) lent support to a thought which—though not publicly aired during the Symposium—had not been absent from some of the private discussions: it was at least not unimaginable that some aspects of nature might be so complex as to be beyond the comprehension of man, even man equipped with all his panoply of ''peripherals'' including those of *informatique* in its fullest sense.

In contrast to this, and certainly encouraging for anyone beginning to sense incipient intellectual indigestion, Professor Harold Linstone (Futures Research Institute, Portland, Oregon) drew attention to a typical case—an exercise in urban planning in Athens—where the application of ''common sense'' had produced greater efficiency in energy use than had a highly intellectual ''first principles'' study. However, the Professor recognised that this was not a condemnation of ''the scientific method'' but rather a demonstration that a problem could often prove a great deal more complex than it might seem on first view. Was this not, after all, support for Professor Prigogine?

Throughout the Paris Symposium, there had been a general tendency to discuss communication in human activity on the

*''Entities must not be multiplied unnecessarily''—i.e. the simplest explanation compatible with observed phenomena should always be adopted. The *dictum* is due to the English Franciscan monk William of Occam (1300–50).

assumption that it was a beneficial instrument normally used beneficially (i.e. for the "best good"). It was left to Professor Luigi Mendia of Naples University to point out that modern communications systems offered important possibilities for "manipulating" information made available to the public, and in some cases even to those with intellectual and/or public authority. This aspect—the deliberate misuse of science and technology—had certainly been in the minds of many participants, particularly during the discussions of "high-level" machine handling of information.

At least for the moment, however, there was no suggestion that the manipulation could result from any but human stimulation. The idea that a *Deus ex Machina* might itself engage in such an activity was not considered, until the next DISCOVERIES Symposium, in Stockholm in 1979.

CHAPTER 4

Automated Information: The Fourth Discoveries Symposium, Stockholm, 1979

H. W. LAWSON

Introduction

The fourth and final symposium in the Discoveries series was held 13–17 August 1979 at the Grand Hotel, Saltsjöbaden in suburban Stockholm. This pleasant seaside site provided a superb environment for the symposium. In this chapter, we first consider the goals and organisational aspects of the Stockholm Symposium in order to provide the reader with an understanding of how the scientific committee designed and developed the Symposium structure. Next, we consider a summary of the opening session which was to provide a starting point for the scientific activities. To indicate the main "results" of the Symposium, a section is dedicated to the summaries by the respective chairmen of the Symposium component parts. The chapter terminates with the most important concluding remarks made at the closing session. Further documentation of the Symposium appears in the form of a proceedings and the descriptive accounts of Mr. Bruce Adkins, Symposium rapporteur.

Symposium Goals and Organisation

The scientific committee, under the chairmanship of Professor Gunnar Hambraeus of the Royal Swedish Academy of

Engineering Sciences and vice-chairmanship of Professor Shuhei Aida of the University of Electro-communications and the Honda Foundation, in organising the Discoveries Symposium for Stockholm, decided upon the theme of "Man and Society—Automated Information Processing". The ordering within the title was deemed to be important; namely, to emphasise the point that it is man and society that are to be the beneficiaries of the automated information processing technology and *not* that automated information processing be the captor and influencer of man and society. This emphasis is directly in line with the goals of the Honda Foundation in sponsoring the Discoveries series; namely, the "Humane Use of Human Ideas".

The major goal of the Symposium was to bring together a cross-section of experts and laymen from a variety of disciplines and through tone-setting presentations and discussion shed light upon the major issues of this important subject area. At an early stage, the scientific committee agreed, that in order to help to keep the discussions down to earth, an actual application of automated information processing technology should be presented in the form of a case study. The subject of "Automated Information Processing in Medical Care" was deemed to be an appropriate area where, hopefully, man and society can benefit from the use of automated information processing technology. Professor Sixten Abrahamsson of the University of Gothenburg, due to his active role in planning and implementing medical information systems, was asked to chair the case study which was to be given as an interwoven part of the subject areas of each day.

Having decided upon the case study area, the scientific committee set about organising the details of the Symposium. An opening session, with appropriate formal introductions, a tone-setting keynote speech and case study introduction plus a closing session to summarise the results of the Stockholm meeting and the Discoveries series, were to be definite anchor points of the program. Professor Gunnar Hambraeus was selected to chair these sessions. The determination of the major subject areas for the three middle session days was rather straightforward. Firstly, a session to provide information, particularly for the laymen at the Symposium, concerning historical aspects, current state of the art and future trends of *The Machines* for which Computer Science Professor Erik Sandewall of Linköpings University was selected as chairman. Sociologist and Rector Emeritus of Uppsala University, Professor Torgny Segerstedt, was selected to chair the session related to assessing the relationships of *The Individual* and automated information processing. Finally, for the session related to *Society and Computers*, Political Science Professor Gunnar Hecksher was selected as chairman.

The scientific committee developed the Symposium program by selecting appropriate people to present tone-setting lectures, including the distinguished opening session keynote speaker Heinz Zemanek, Professor at the University of Vienna and Fellow of the International Business Machines Corporation. Professor Zemanek, in his previous capacity as President of the International Federation of Information Processing Societies (IFIP) has been instrumental in bringing attention to the importance of discussing the impacts of automated information processing upon society. Therefore, Professor Zemanek's presentation entitled "The Impact of Information Processing Upon Society" provided an authoritative start to the scientific program. Professor Reikichi Shirane of the Telecommunications Science Foundation, Tokyo, Professor Sidney Michaelson of the University of Edinburgh and Professor Edward Fredkin of the Massachusetts Institute of Technology were selected to present the lectures related to the historical aspects, current state of the art and the future trends of the machines. For the tone-setting lectures of the sessions related to the individual and society and computers, Professor Harold Linstone, Portland State University and Professor Klaus Lenk of the University of Odenburg, West Germany were selected.

The strategy in selecting the remaining participants was to obtain a mixture of previous Discoveries participants and new participants that would provide expertise in the Symposium subject area. The Symposium participants (thirty-six in number) including the scientific committee and lecturers are listed elsewhere. The distribution by country is as follows: Sweden, the host country (9), Japan (9), United States of America (5), Great Britain (3), Austria (2), Italy (2) and France, West Germany, India, Kenya, Norway and Yugoslavia (1 each). From the participants, several were selected to provide short prepared discussion introductions on special topics as published in the symposium schedule.

Review of the Opening Session

The Stockholm Symposium was officially opened by His Majesty King Carl Gustav of Sweden. His Majesty emphasised the importance of the subject area by stating that "computers are here to stay and that we must be aware of both the positive and negative effects of the presence of computers". Session chairman, Professor Gunnar Hambraeus, after welcoming the participants on behalf of the Royal Swedish Academy for Engineering Sciences, invited Japan's ambassador to Sweden, H. E. Masahisa Takigawa to read a special message from the Japanese Prime Minister. Foreign Minister Hans Blix, representing the Swedish government, pointed to the importance of the merger of telecommunications and computer technologies and the potential positive and negative effects of this union. Dr. Soichiro Honda then, in addressing the participants and opening session guests, vowed to devote the remainder of his life to the Discoveries Idea and Project.

In his keynote speech, Professor Heinz Zemanek emphasised the transitional problems in the introduction of automated information processing in the short run but was optimistic for long-range prospects. Effects upon labor and employment forms is unavoidable. The computer terminal is destined to be a factor in human communication and will certainly be viewed with mixed reactions.

Professor Zemanek provided an historical account of the evolution of computer languages. Further, he made a statement, to be challenged by several participants, that "although a computer might change, or destroy information, it could not generate any new information with more basic content than that fed to it''. Professor Zemanek speculated that electronic voting systems for the citizenry of a country, although not attempted thus far, will be attempted in the future. In his conclusion, the Professor stated that ''our fate will remain in our own hands, because computers will return decision-making to us'' (a thought-provoking statement).

The opening session terminated with the case study introduction by Professor Sixten Abrahamsson. The results of the case study are summarised in the next section.

Summaries of the Case Study and Session Chairmen

The sessions of the following three days provided for stimulating presentations by the tone-setting lecturers as well as lively discussions in the meetings and in smaller groups during pauses. Since the information transmission was largely ''simplex'' (one-direction) during the first day, the discussions were most intense during the days related to the individual and society and computers. The closing session of the Stockholm Symposium was held in the Swedish House of Parliament in central Stockholm. Professor Eduardo Caianiello, participant at all four Discoveries symposia, first presented a summary of the previous symposia in Tokyo, Rome and Paris. Following Professor Caianiello's presentation, the case study and session chairmen presented their summaries which are presented here, in edited and in certain cases complemented form, to provide the reader with the main ''results'' of the three working session days of the Stockholm Symposium.

Automated information processing in medical care

The demand for health care is increasing all over the world and no one seems to believe that a saturation can be reached. Health care represents a labour intensive industry which now faces severe shortages in both funding and personnel. In Sweden, the costs are rapidly approaching 10% of the GNP. In this situation, automated information processing has been introduced as a means of reducing the costs and/or increasing the use of available resources. This process, which started at too early a stage some 20 years ago, has now matured and will accelerate considerably during the 1980s due to the rapid advances in microelectronics. Such a development will radically influence both man and society and it was therefore highly appropriate to illustrate some of the effects as a medical case study throughout this symposium.

Some of the special problems when designing medical information systems have been identified. The key issue is to arrive at a solution that is *acceptable* by the very different user categories; that is physicians, nurses, medical secretaries, administrators, politicians and patients. Their needs are so widely different that the final design necessarily represents a *compromise.* If this is not made quite clear to all users, the system will never be accepted. An alternative with separate systems for the different user categories would lead to unrealistic costs and would not give the total picture of the health care delivery system necessary for control and planning functions.

At present, most functions in health care can be, and have been, automated. Some are cost-effective such as systems for the clinical laboratories. The use of microcomputers will further improve the performance of such systems. The number of special stand-alone equipment requiring little or no attention by computer people will increase drastically during the 1980s. Several subsystems again are not cost-effective but are operated for other reasons such as being a complement to a cost-effective routine at marginal extra cost.

In order to achieve a better resource utilisation, better planning and improved care, the various subsystems must be made to work in intimate interplay. An integrated medical information system (a concept of the 1960s) must therefore be constructed with at least the basic functions of health care included. Though we know by now the human problems involved, the task to make such a system an integral part of the routine health care operations is by no means easy.

Computer simulation can be used with advantage to forecast medical demands and resource consumption. In the first case study presentation, Professor Kazuhiko Atsumi revealed that the morbidity of degenerative diseases, including cardiovascular diseases and malignant cancer, follows a similar pattern in different countries. Thus, when various individual population structures are taken into account, the total number of people needing care and the associated costs, as well as the death rates can be estimated with a high degree of reliability on the international level.

The model indicates to decision-makers which actions are necessary in order to fulfill a certain determined policy for health care operations. Professor Atsumi furthermore made the point that "health care becomes a science by the introduction of computers".

The comprehensive medical information system of Stockholm County was used as a base for the two other case study presentations. The system covers the entire medical region with 1.5 million inhabitants and records data on each patient visit. The total information is instantly available over some 350 terminals to all authorised users.

Docent Peterson, in his first presentation on the third day, introduced a relatively new feature of the system; namely, care programs for cancer. All physicians of the region have agreed not only on the way of treating various forms of cancer but also on all information that should be stored in each case. The number of cases treated in the same way thereby becomes relatively large which allows a follow-up to be made earlier than before. There is also the possibility to divide the patients into different treatment groups and then compare the results. This routine is of obvious benefit to the patient but also to the health care authorities by reducing the number of expensive hospital days.

The final case study presentation on the fourth Symposium day (also by Docent Peterson) dealt with the total registration of all inpatient visits and its effect on the planning function. The information collected since 1969 can be used to trace changes in the panorama of diseases. Harmful effects on the individual by the environment and of social attitudes, for instance towards drugs and alcohol, can be detected at the earliest possible time.

The consumption of health care resources can be compared to the expected one calculated, for instance, with Professor Atsumi's model. Areas of high overconsumption exist in the County. This can most often be explained by socio-economic factors; whereas, the cause in some cases is unknown. The authorities now have the facts and should take measures to correct the situation.

Projections of the number of beds needed, say 10 years from now, can easily be calculated from the data files. This is, of course, of crucial importance since the construction of new hospitals takes considerable time.

During the Symposium many worries were expressed over storage of data on computers and privacy. The very effective security system of the County was demonstrated. It is based on an individual password which is changed every quarter year. From the password, the computer identifies the user and checks that he may access the requested data. In addition, a control is made that the user is at an expected terminal, say in his own clinic. All transactions are logged and regular checks are made on what users have been doing with the system or who has been requesting particular data. Should someone try to break the code, the terminal is physically closed after the fourth attempt.

The system has never failed and the computer files must be regarded as more secure than the manual medical records floating about in the hospital.

The Symposium never proposed any resolutions concerning automated information processing in health care. The case studies, however, served the intended purpose of providing realistic down-to-earth contacts for the sometimes high-flying thoughts of the delegates.

The machines

The purpose of the second day of the symposium was to review the world of information processing machines in dimensions of time (discussing the past, present, and future of automatic tools), and space (making comparisons between uses of computers in different parts of the world), to provide the necessary background of technological facts and predictions for the discussions in subsequent days.

The session began with a systematic exposition of the history, nature and impact of automation and automated information processing. Presented by Professor Reikichi Shirane, President of the Telecommunications Science Foundation in Tokyo, this set the scene by recalling the origins of large-scale automatic processing in the US car industry some half a century ago; the subsequent development of multiple servoloop systems (initially mainly for military purposes in World War II); and the post-war expansion of such concepts into cybernetic automation or "cybernation".

Professor Shirane traced the effects of these various stages in man's increasing use of sophisticated control arrangements, not only in industrial production but also in fields such as agriculture and mining, transport and communications, accountancy, banking, and in "non-industrial social systems" such as management and administration, education and medical care. He pointed out how these applications had been made possible and worthwhile by developments in the technical performance of the computers used—notably increased operation speeds and computing capacities, increased memory capacities and reliability, and greatly reduced costs. However, he did not suggest that any of these attributes made the machine more than an important *extension* of the information processing capabilities of the men in control; and he concluded with some challenging remarks on machine/man relationships in Japan and the impact of the ever more ubiquitous computer on the Japanese social scene.

Professor Shirane was followed by Professor Sidney Michaelson (Edinburgh University) whose presentation of "The State of the Art in Information Processing" was a compendium of facts and figures illustrating the remarkable progress of computing machines over the past quarter century. To cite just one example of this progress, he gave some details of a typical "quite large machine" of the early 1950s: twenty to thirty high-speed registers and a file store of a few thousand numbers. Today he said (or rather, in view of the speed of current development, yesterday) children can buy as *toys*, devices of very small physical size which have a thousand times more immediate working storage and a thousand times the backing storage. Professor Michaelson also pointed out the broad spectrum in size of computers, from the very small to the very big. The concept of a general purpose computer is becoming less and less appropriate, as computer equipment is specialised to various tasks.

The discussion about effects of computers on individuals and society had been scheduled for subsequent days, but started early, on the second day. Parts of Professor Michaelson's speech, as well as subsequent statements by Professors Linstone, de Sola Pool, Finer, and others focused upon the effect of computers on employment. This was to become one of the main topics of interest during the following days.

In the afternoon, Professor Edward Fredkin of the Massachusetts Institute of Technology spoke on "Long-range Prospects for Intelligence in Information Processing Systems". Professor Fredkin observed that the computer is already being used as a prosthesis for human thought in certain areas, and the number of such areas is likely to increase as advances in hardware and software technology make it possible to program more complex information-processing tasks for computers. In the development of the technology, certain facilities such as an internal autonomy of the software system, and an ability for it to modify itself, will make it appropriate to call such systems "intelligent".

Professor Fredkin argued, in this context, that (1) the prerequisites for intelligent computer systems are already coming into existence in the form of cheap LSI technology, widespread storage of information on computer media, etc (2) to develop the technology of intelligent machines, one single breakthrough is needed, and may be performed even in a small laboratory with very modest equipment. When the breakthrough is done, an over-flood of rapid development is likely to follow. Furthermore, the longer the time lasts until the breakthrough comes, the quicker will be the subsequent developments; (3) like all other new and complex machinery, intelligent computers will certainly contain design mistakes ("bugs") when they first appear, and may therefore be dangerous.

Professor Fredkin therefore concluded, first that a quick development of intelligent computer systems would be desirable to guarantee and even control the use of this technology, secondly that when such systems first appear, they should be put into quarantine, and thirdly that an international agency should be formed for research, development, and control over intelligent computer systems.

The discussion following Professor Fredkin's talk treated both his analysis of technical possibilities and political proposals. There was disagreement about the concept of a single breakthrough—some speakers believed rather in a gradual development which would be easier to control. The political desirability, as well as feasibility of the proposed research agency, was also controversial, but no alternative solution for handling the problem was proposed.

In summary, the first day was mostly used for providing different points of view on information technology. Much of the discussion that started in a small scale even on the first day, was to continue on the following days.

One issue which would recur repeatedly during the Symposium, became very clear even on the second day; some participants believed that the effects of computers in society can best be studied by extrapolation from previous use of computers, and their effects so far; various other participants felt that such extrapolation would be very unreliable, and often misleading, and that one must instead look at the properties of computer technology as such in order to get an understanding of their likely effects in the future. Since they covered both past developments, present state in several countries and long-range prospects, the speakers during the first day provided an ample background for both approaches to studying the effects of information technology on individual and society during the DISCOVERIES Symposium.

The individual

The third day was devoted to a discussion about the impact of electronic technology on the life of individual man, on his family life, his education and his work. It was generally felt that all sectors of human life would be influenced. I can only mention a few consequences, or possible consequences, of new computers and terminals.

It was first of all repeatedly pointed out that the new machines were getting cheaper and cheaper and could before long attain a wide distribution. The computers are however in a way selective: only certain traits of reality are treated by them. As Professor Harold Linstone, the tone-setting lecturer, put it: "The computer requires a mathematical model, but such formalism is unable to represent the relevant properties of organisational and individual perspectives". Because of this aspect, computers are creating new realities. We may say that the specific *data language* structures our experience of reality. As computers and terminals seem to be very common, a consequence seems to be that we all must learn their language. If we do not understand that specific language, the machines may get a power of their own. To understand the technical language may be said to be a democratic demand which is particularly felt in the workforce, but not as important for all individuals.

The reality created by computers includes our way of looking upon other human beings. There is, as we all know, a risk that we are intruding on their privacy and their personal integrity. But there is evidently also a danger that the electronic machines only will underline our rational modes of thought and neglect our emotional and intuitive capacity. For that reason we must remember that *social interaction* is an important factor in human development. One new technique may diminish our face-to-face relation as a large amount of communication can take place via computers. When analysing the consequences of the new technology, we cannot neglect the energy problem. The important question is: if the new technology requires more or less *energy* than the conventional techniques? If it requires less energy, we will probably be forced to accept it whether we like its consequences or not.

It was furthermore evident from our discussions that when trying to predict the implications of computers on the life of man, we had to take into consideration what is possible from (a) a technical point of view, (b) a psychological point of view, (c) a political point of view and (d) an economical point of view. It seems likely that the real resistance toward electronic technology will be of a psychological and political nature.

We also learned, as often before, that the problems may seem different if you look upon them from the Western point of view or from the developing nations' point of view. We are apt to forget that problems may be quite different from different viewpoints.

When we discussed the new job-situation, it was pointed out that a significant amount of work in the future (if not already) can be carried out at home, as you can obtain enough information from your home terminal. The work mentioned was mostly white collar work. But if working at home becomes common, it may solidify family ties.

The question remains: could ordinary industrial work be decentralised in the same way as white collar work to small units at the same time as a centralised source of information and control directed the production program? If that could be possible, we would have a new kind of cottage industry or village industry. The cottage industry implies a new kind of social life and consequently has enormous implications on labor life and must be of great importance for labor-unions. It was stressed, at our discussions, that these changes must not be carried through without information and concern from the persons affected. The fear is, of course, the disappearance of jobs and an increasing unemployment. The problem is how much liberty of choice the energy shortage will admit to us. I am sure there will be a strong psychological resistance and probably a strong political resistance. There is a general feeling that employees must not only be spectators but actors and controllers as well.

I hope the psychological and political resistance will not be so strong that we do not observe that the new machines may relieve the individual worker of dangerous jobs such as handling poisonous material. Robots can take over that kind of non-human job. With regard to our daily life in our homes, it was said that the distribution of consumer goods could be drastically changed. Our daily needs could be programmed and goods delivered when we were running short of them. It does mean a decrease of social contacts; however, it may solve the problems for old people and of out-of home working parents. May I add that robot-servants to do our house-hold jobs were also declared possible. The computers and terminals make decentralisation possible. That may influence the power structure in many ways. Some of them were mentioned during Professor Gunnar Heckscher's day on society and computers. We did not discuss however, if that meant that the family would have a stronger position in future society or if new primary groups would emerge, namely, groups which interacted around the new machines.

As I have already said, much fear was expressed that the new machines might create unemployment. I am sure we will have an increasing leisure time whether we want it or not. A considerable part of that leisure time will, however, be consumed by retraining men and learning new skills. Speakers meant that electronic devices could be very helpful in the vocational retraining of people who have lost their jobs or had to adjust themselves to the changed condition of their old jobs. It was also pointed out that the new machines gave everybody a chance of *life-long education*. So the situation seems to be that electronic technology may kill jobs but, at the same time, it helps to re-educate people for new jobs. One of those new goals of education will be to understand the language and power of our new instruments. But even if it is true that a rather large part of the leisure time will be consumed by adult education, there will still be a large amount of time at our disposal. There is no doubt that the new machines can distribute culture goods in quite an effective way. If we have new primary groups and revived villages they will be the consumer of our culture goods. I regard that as very important. But I am sure that we have to educate indviduals to understand culture values. That can probably *not* be done by computers. I have talked about the danger of individuals being isolated or that they will be without a chance of human interaction. One speaker said that "we now have evidence that *certain forms* of social interaction, which help the child to take a perspective other than his or her own on what is happening out there, are a powerful ingredient of the development of abilities". It is evident that computers can be instruments of education, but that they never can replace human beings. It is still so, that human beings are being human only in interaction with other human beings.

Society and computers

The discussions during the fourth day, held under the title "Society and Computers" were lively (with no less than 42 interventions by 21 different participants) and covered a wide field. Any selection of the points discussed, consequently, must be subjective and even arbitrary, but I shall mention some of them. If I fail to mention by name the participants making the different points, it is because very frequently the same question was analysed by more than one of the speakers. It was characteristic that there was real give and take all through the discussions which we, in the West, believe to be the best way of making discoveries.

(1) The tone-setting speaker, Professor Klaus Lenk, posed the basic question: Is AIP (Automated Information Processing) going to remain a tool, or does the technology tend to develop so independently that it may become our master? Although it was pointed out that both may be true to some extent, most speakers agreed that, for the foreseeable future, it will remain no more than a tool and that we thus carry a heavy responsibility in deciding how it is going to be used.

(2) Democracy has learned how to deal with new technologies. But beneficial use of AIP requires education not only of decision-makers, but also of ordinary citizens concerning its character. The technological experts must provide the information necessary for this purpose.

(3) International control could be desirable, as Professor Shuhei Aida pointed out, but it is hardly a practical possibility. To some extent, its place can be taken by wide publicity.
(4) The industrialised societies must make serious efforts to take into account the interests of developing countries. Technology should be adapted to their needs and it would be a great mistake to believe that what is viable in highly sophisticated Western societies could be transferred without important modifications to emerging societies with different cultural backgrounds.
(5) In any case, the use of AIP involves real political problems, where both different interests and different ideologies may appear to be in conflict with one another.
(6) As far as industrialised societies are concerned, it is likely that AIP will initially enforce bureaucratic power and cause centralisation. However, its present effects do not necessarily indicate what the ultimate results will be. New technologies tend to favour central governmental power as long as they remain expensive, but when they become inexpensive, the opposite may be true. Moreover, AIP can already now be used as an instrument of subversion, and small groups may be able to disrupt the power structure, if it is dependent upon centralised machinery of strategically important points. Also, widespread use of AIP is likely to enhance the visibility of governmental and administrative procedures in relation to ordinary citizens.
(7) Fear of the administrative consequences of AIP is motivated by the assumption that administrative agencies are malevolent. However, in democratic societies this is not true. On the whole, they try, if with varying measures of success, to realize the objectives decided upon by democratic process.
(8) The fear that AIP may be a threat to privacy and to the integrity of private individuals is not warranted, if necessary precautions are taken. In the case study, an example was given of how such precautions work in the

health services of Stockholm county.

(9) Much controversy appeared in the discussion of how AIP is likely to affect the labor market. No agreement could be reached as to whether or not it can be expected to result, on balance, in loss of jobs. However, there seemed to be agreement that it would be unreasonable even to attempt to fight the introduction of this new technology. At most, employees' organisations should be given power to participate in its development and to postpone, for a short time, such changes as might have dangerous effects on the labor market. Many problems could also be solved by active labor market policies for the creation of employment in new fields.

In conclusion, it could be said that our discussions on "Society and Computers" identified a great number of problems and indicated that most of these are far more complex than is generally assumed. We cannot claim to have found definitive solutions—but this was only to be expected.

Note: In subsequent correspondence, Diplom Engineer Fred Margulies from the University of Vienna, an active participant in labor union activities in Austria, requested the following text for Professor Hechsher's point (9) to more adequately reflect the discussions from the labor point of view.

(9) Much controversy appeared in the discussion on how AIP is likely to affect the labor market. No agreement could be reached as to whether or not it can be expected to result, on balance, in loss of jobs. However, there seemed to be agreement that it would be unreasonable even to attempt to fight the introduction of this new technology. At the same time, there seemed to be agreement that things should not be allowed to drift along. Rather than trying to predict the future, we should shape it; rather than speculating about protecting ourselves *against* the computer, we should create conditions which will allow us to draw all the benefits *from* the computer. Employees' organisations should be given power to participate in decisions concerning technological

and economic development, because solutions for these problems cannot be found *for* people, but only *with* people. Many problems could also be solved by active labor market policies for the creation of employment in new fields.

Concluding Remarks

During the closing session, the all important "DISCOVERIES Declaration" was made heralding the introduction of the HONDA PRIZE in Eco-Technology. In closing the session, Professor Gunnar Hambraeus gave the following account of the achievements of the Stockholm symposium.

First, technology had been shown a true fruit of the Tree of Knowledge, to be used by man for good or for evil.
Second, the latest developments in automated information processing and related fields were correctly considered as no less significant than the Industrial Revolution.
Third, these developments engaged the responsibility of all concerned: from the scientists to the politicians. There was no excuse for avoiding this responsibility—in the case of the scientist to explain, in the case of the politician to understand. Without understanding there could be no soundly-based decisions.

Postscript

As scientific secretary for this final DISCOVERIES Symposium, I would like to express my personal appreciation for all cooperation and companionship of my colleagues from the scientific committee, the organisation committee and the participants. The Stockholm Symposium provided a valuable experience for all and our appreciation is given to Ambassador Takeso Shimoda, Honda Foundation Director and to Dr. Soichiro Honda for making the Symposium possible.

Harold W. Lawson Jr
Professor, Linköpings University
Scientific Secretary

CHAPTER 5

Evolution and Revolution

E. R. CAIANIELLO

THE problem of the age of the Universe is crucial in today's natural philosophy; present evidence points to a pulsating Universe, that explodes from point-size into an expanding whirl of matter and radiation, to recontract, to again re-explode. Physics goes here beyond imagination; the highest authority on the subject, J. A. Wheeler*, compares this impasse with the failure of classical physics that gave way to the Quantum Theory (Vedic philosophers would, though, happily agree to this picture: even their estimate of the duration of a cycle coincides to within a factor of ten with that of modern astrophysics). Let me just say, that nobody would dare claim today to any degree of certainty on this subject.

The situation was very different not so long ago. Two years after the printing of the "Dialogue Concerning the Two Principal Systems of the World" by Galileo†, in 1634, the "Prince of

*__Wheeler, John Archibald (born 1911)__ American physicist. When Otto Hahn of Germany *et al.* discovered nuclear fission in 1938, Wheeler proposed the theoretical method to explain this phenomenon together with Danish physicist Niels Böhr (1885–1962). Also, in connection with the problem of describing matter in the general theory of relativity, succeeded in giving an interpretation to mass and electric charge in 1957, by expanding German physicist Albert Einstein's (1879–1955) way of thinking into the unified theory of field and using harmonic integration.

†**Galileo Galilei (1564–1642)** Italian mathematician and scientist of the late Renaissance period. One of the pioneers of modern physics. Discovered the "isochronism of the pendulum" in 1582 at the age of 18, and performed an experiment on free falling bodies at the leaning tower of Pisa in 1589, thereby proving the fallacy of the Aristotelian doctrine, then considered to be the absolute truth.

In 1609 invented the telescope, utilising it for astronomical observation. In the following year discovered Jupiter's satellites as well as maculae on the sun and the moon, in addition to making numerous other astronomical findings that served to prove the correctness of Copernicus' heliocentric theory.

chronology'', Jacques Auzolles de la Peyre, was in a position to state that the age of the world was exactly 5954 years. The Bible permitted precise mathematical computations; creation had taken place on the autumn equinox, a Wednesday.

How has such a change in knowledge and perspective occurred? How has Man changed his familiar cozy small world into an infinite expanse, his certainties into anxieties, to find himself again delved into another, perhaps bigger cultural revolution, entered with easy optimism and soon found likely to portend unspeakable disaster? To speak of cultural change, I think no better situation could be proposed for illustration and thought.

Two cautions must be kept in mind before venturing any analysis of the impact of innovation on culture. The first concerns the use of language. Though this need hardly be stressed in the country of Zen, *words are traps*, they easily become the masters rather than the tools of thought. The greatest obstacle against the acceptance of Relativity has been that ''time'' and ''space'' mean something entirely different in physics and in common parlance, and yet it is the latter that sets up mental patterns: wasn't it ''absurd'' that different observers might not agree on events being simultaneous? The examples are innumerable: the *De revolutionibus Orbium Coelestium* by Copernicus* expounded in 1543 A.D. the very same heliocentric doctrine advanced by Aristarkhos† of Samos in 280 B.C.; but the cultural pattern imposed by words had drastically changed by then, ''heavy'' and ''light'' did not necessarily go with ''earth'' and ''fire'', and,

***Copernicus, Nicolaus (1473–1543)** Polish astronomer. Proponent of the heliocentric doctrine which he expounded in ''De revolutionibus orbium coelestium'' (1543), to overthrow the geocentric doctrine advocated since the time of Greek astronomer Ptolemy (Ptolemaios Klaudios, second century A.D.). In so doing, opened the way to modern astronomy.

†**Aristarkhos (ca. 301–230 B.C.)** Greek astronomer. Known as the originator of mathematical geography and geodesy. Said to be the earliest to believe that the earth undergoes revolution around its axis and round the sun simultaneously, Aristarkhos gave a strong hint to Copernicus in his formulation of the heliocentric doctrine later in history.

though the sun was still "fire", it was no longer "obvious" and "natural" that the earth should be as far down as possible, here, and the sun float lightly up and around, as with the Ptolemaic system: the main, hidden objection had ceased to exist. Such is the power of words, and of the associations that are formed of them, beyond our awareness, in any culture, including of course the one that has us. All conditionings embedded into words are the first we have to unlearn if we are to think of the "next" culture.

The other caution has to do with our predictive powers: they are very, very limited, for at least two reasons. Statistics give an "average value", but we live always in a "fluctuation" around that average, and I can hardly be interested in the prediction of a value of which the same statistics prove that it has vanishing probability of being true here and now (I am quoting Norbert Wiener,* who estimated the duration of a fluctuation in social predictions to be no less than a human lifetime). An even more forceful reason is that culture, science and technology form an adaptive system, which changes its structure as a consequence of its evolution (of this I shall have to say more), but mostly because of unforeseeable discoveries (of America, of atomic power, of computers etc.): what sane man would venture predictions about a system for a time when its structure will have changed in a way and by an amount certainly unknown? Yet we are far from powerless: if we cannot predict a distant future, we do determine it in every way with our actions.

A measure of the magnitude of the change that took place when the Middle Ages gave way to modern science is given by the cited problem of the age of the Universe; but there is much more than

***Wiener, Nobert (1894–1964)** American mathematician. Founder of the science of cybernetics. His formulation of cybernetics was motivated by his experience in the design of computers as well as his joint research with neurophysiologists.

Based on the recognition that a functional similarity exists between the construction of computers and the structure of the nervous system in living organisms, cybernetics attempts to explain such a similarity through a unified mathematical theory. Not only does it serve as the starting point for today's theory of man as a machine, but also constitutes one of the theoretical bases for automation technology. Author of "Cybernetics: or, Control and Communication in the Animal and the Machine," 1948.

that. To really size up the issue one must make the almost impossible attempt to forget what is obvious and natural *now*, and to readjust to what was obvious and natural *before* the revolution took place. We must get into a world that, besides being less than 6000 years old, was also very small: it amounted to no more than the Earth, rooted at the centre of everything and meant only for Man, with a pageantry of lights revolving around it, in geometric circular perfections, possibly pushed by angels. The very concept of "space" was far from ours: Euclid* treats "objects", only with Descartes these became "portions of space" (the concept of "infinite" was surreptitiously introduced by the Jesuits of Coimbra with the notion that, finite as the world might be, it still had to be contained in God, who possessed the metaphysical attribute of infinity; slowly the divinity disappeared and the attribute remained). Science—its quintessence geometry—was abstract speculation, worthy of intellectual respect because "pure" and not to be polluted by contact with the "mechanical arts", which obliged to use instruments and had practical, therefore less meritorious purposes (the major exceptions were Archimedes among men, astronomy among sciences: but this served the practical aims of navigation and horoscope casting, so Ptolemy† could properly describe in his Almagest the astrolabe and other instruments). Of Plato‡ and Aristotle§ (and Pythagoras,

***Euclid (330?–275? B.C.)** Founder of geometry. Through his writing "Stoicheia", Euclid perfected geometry for the first time in history as a superb axiomatic system of deductive reasoning. Euclid's geometry set the norm as the only geometric system, until N.I. Lobachevskii's (1793–1856) non-Euclidean system made its appearance.

†**Ptolemy (Ptolemaios Klaudios; second century A.D.)** Greek astronomer and geographer. His principal astronomical writing "Megale syntax" ("Almagest" in Arabic) was the comprehensive compilation of astronomy practised in Greece and Alexandria.

‡**Plato (427–347 B.C.)** Greek philosopher and follower of Socrates. In his theory of human soul versus the body, Plato asserted the immortality of the soul and taught that Idea, the timeless essence or universal which is the archetype of an individual existent as sensed by the soul, is the truly immutable real existence.

§**Aristotle (384–322 B.C.)** Greek philosopher. Studied at the Academy taught by Plato. While Plato considered Idea as being transcendental, Aristotle on the contrary believed form (Eidos) as being the essence inherent to matter (Hyle), and explained all formative processes as processes by which matter actualises form.

Philolaos*) I need not speak here: like with Yin and Yang for Chinese things, it has always been a favorite game for Westerners to trace their thoughts back to them. The true age of logic was that of Scholasticism, St. Thomas Aquinas brought Aristotelianism to diamond perfection: as with the Coran for Muslims, what was in it was true, what was not, false.

The crucial point, in my opinion, is that this whole body of doctrines, that lasted over two thousand years (in places is not dead yet), clumsy and absurd as it may seem to us *now*, was in fact a perfectly coherent, majestic and faultless system of thought. It was in excellent agreement with all known facts; furthermore, it responded beautifully to the claims of common sense: the sun does move around the earth, heavy bodies fall because what is heavy tends to go down, force is proportional to velocity as everybody who tries to push a cart knows; astronomy —the geometry of luminous, therefore pure entities—has nothing to do with physics, the science of what grows and decays, rises and falls, is impure; and what use can a system of thought be, if it does not explain everything at the same time, animals and plants and the cosmos? Why stoop at so trivial a question, how does a falling stone fall, inch by inch?

Newton's† *Philosophiae Naturalis Principis Mathematica* appeared in 1687; one might say—comparing with the 1543 of Copernicus's treatise—that in about 150 years the full revolution had been accomplished, events more important for mankind than all wars had taken place. One innovation caused all this: the inductive or experimental method, that we call Galilean, replaced recourse to authority with plain, persevering observation

***Philolaus (420? B.C.)** Greek philosopher of the Pythagorean school. His theory on celestial bodies said that the earth, endowed with a spherical shape, was suspended in space and furthermore, that it revolved around the fire (Hestia) at the center of the universe, as did other planets, fixed stars as well as the sun, in accordance with a definite mathematical law.

†Newton, Isaac (1643–1727) British mathematician, physicist and astronomer famous for his three great discoveries: resolution of light, the law of gravitation, and differential and integral calculus. His great accomplishments are contained in *"Principia mathematica philosphiae naturalis"* (1686–1687), which encompasses his entire achievements in the field of mechanics. This mathematical and physical treatise became the foundation for the study of mechanics for the next 200 years.

and measurement; attention was fixed on simple and reproducible phenomena (such as the falling of stones) rather than on complex wholes; people stopped asking "why" and started searching "how"; the "explanation" of facts, not sought any more in names and final causes, was replaced with their "description", for which a new language was found necessary. Hear Galileo ("The Assayer"): "Philosophy is written in this most large of books that stays continually open to our eyes (I say the Universe), but it cannot be comprehended if the language of it is not learned before and the characters in which it is written are not known. It is written in mathematical language, its characters are triangles, circles and other geometrical figures; without such means it is humanly impossible to understand a single word of it, without them there is only inane wandering through a dark labyrinth".

The path of innovation, I propose to state, involves both evolution and revolution. Let me first illustrate this concept with a mention of Leonardo* and Galileo. We have no room here to describe, or even name, the causes of Italian Renaissance. To set the stage, think of wars storming all over the country; of the rediscovery of ancient art and philosophy; of the fall of Constantinople to the Turks in 1453, that flooded the Italian courts with learned and haughty refugees; of the ennoblement of skilled labor (no longer a debasing activity: Florentine artisans could invent lenses in the XIII century because they had *not* gone to school to learn that lenses alter truth to the eye); of the appearance, near the Prince, together with the Physician and the Astrologer, of the Engineer (whose skills usually went from the military arts to much higher attainments: Leonardo, offering his

*__Leonardo da Vinci (1450–1519)__ Artist, scientist as well as engineer, and a great master of Italian Renaissance art. At the same time a pioneer of modern scientific thought in a tradition starting with Galileo Galilei. The scientific philosophy found in da Vinci's voluminous manuscripts emphasise the inseparable bond existing between theory and experiment as well as the application of mathematical techniques to the study of natural science, and thus contains a novel methodological discovery that breaks down the medieval and Aristotelian view of Nature.

services to Ludovico Sforza*, Duke of Milan, listed in 1483 ten skills of which he deemed himself capable: nine of them concerned military engineering). These and many other occurrences were handed down and accreted from century to century: this type of hard historical facts is what I resume under the name "evolution". A single man stands shoulder high amongst all, to incarnate for us the struggle and quest of his time: Leonardo da Vinci. His achievements in every path open to man have become myth; he cannot be classed by any simple standard, his praises of mathematics or his statements on the fall of heavy bodies or on the concept of force have been variously interpreted as the farsighted illuminations of genius, or as abstract, platonic remarks not implemented into concrete action. Him we cannot simply judge, if for no other reason, because the meaning of terms such as "force" has been vague and changing for centuries later, and we cannot know what he meant by them. Nor did he generate a school; he stood aloft but alone. Any normal meter, though, would be improper and irreverent: the only way to approach Leonardo is that of the humble, who revere him as the symbol of his age and look at him for inspiration, not criticism. Leonardo represents an essential element in our discourse: not for what he found, but for what he ever, untiringly, *searched* in all aspects of Nature. Nothing more than the vision of this giant to whom science, technique and art were one and the same thing, struggling against the mists of a dying yet still dominating mentality to snatch out of them glimpses of light toward horizons yet too remote, can make us aware today of the immense difficulties, of the centennial work that was ahead of any who would abandon two thousand years of tradition for a new, truer philosophy of Nature. Only the man Leonardo can give us a

***Sforza, Lodovico (1481–1499)** The Sforzas were an illustrious family under whose rule the city of Milan prospered during the Renaissance. The name Sforza, meaning "Conqueror", was an honorary title accorded to the founding father Muzzo Attendola (1369–1424), who, despite his humble beginnings became Condottori (leader of mercenary troops) and led successful campaigns on behalf of the queen of Naples and the Roman pope. Lodovico Il Moro (1451–1508), who was initially the regent for his nephew eventually wrested the ducal title from him, to become the fifth Sforza to reign over Milan.

measure of what evolution and its inner torment do really mean.

Galileo, on the contrary, must interest us here not as a symbol or myth, but as the actual originator of the new physics and methodology: we are focusing on him to discuss "revolution", which is something drastic and definite, punctualised so to say. We must imagine Galileo at work, see him dispel with strenuous perseverance and lucid clarity the mists from a horizon rendered tangible of a sudden by his own efforts; we want also to realize how things obvious to us (such as Kepler's* non circular orbits, or the fact that we cannot describe "Nature in itself", but only "models of Nature") were for him unthinkable. Copernicus's re-proposal of Aristarchos's system (that he learned in Italy from Archimedes) was or was not accepted by his contemporaries, on academic grounds: Reinhold's† astronomical tables, computed after the Copernican system, were used by supporters and opposers alike. His aim had been to supply aids to navigation; the preface to his book (written by his friend Osiander) declares that it deals with hypotheses useful for computation, not with reality. Whether Copernicus really thought so is a matter for debate; the relevant point for us here is that considering his theory a "model" to get numerical results (a similar stratagem would have spared all persecutions to Galileo, who strongly refused to comply with the suggestion) meant, at that time, *no* philosophical revolution: "model" meant then a tool, mathematical trickery good for practical purposes, not a step toward a deeper knowledge. It is the most recent acquisition of scientific thought that the actual way to improve our knowledge of Nature is through series of models of increasing degree of approximation to reality. This new conviction takes us as far from Galileo as he was from Aristotelianism. It is a matter for meditation that the whole fight of Galileo centered around an issue that is dead for us, though at his time, in his world, it was the necessary and courageous thing to maintain: for, even if we think now otherwise, it was then vital for the

*__Kepler, Johannes (1571–1630)__ A German astronomer. Celebrated for his formulation of Kepler's Law on which the discovery of universal gravitation was based, and which became the eventual key to unravelling the structure of the solar system.

†**Reinhold, Erasmus (1511–53)** German astronomer, Copernicus's disciple, completed a time-table of celestial bodies based on Copernican theory in 1549.

triumph of the new natural philosophy that *either* Ptolemy *or* Copernicus be right: each pretended to a *true* description of Nature itself, not of a model of it; a plurality of descriptions was utterly inconceivable. Such were then, inescapably, the terms of the fight between the principle of authority and the new method of scientific enquiry.

Galileo's travail was titanic, mainly because he refused to compromise: thus it would have saved him much trouble writing in Latin rather than in Italian, but he wanted Italian to awaken the widest audience to the foul prevarication of false doctors who opposed the true and free knowledge of Nature. A study of Galileo's life and ordeals is a rewarding and revealing experience: the adversity of the times, the way he used "thought experiments", his brilliance and impact on scientific thought even when he was wrong. We cannot speak of this here, nor need we describe what Galileo achieved: it is all built in our minds and bones, the way a scientist thinks today is radically different from that preceding Galileo (him we take now of course as the epitome of the whole revolution started and symbolised by him): just recall the brief mention made before of the Aristotelian world. Mathematics has become since Galileo the language of science; Archimedes, Democritus and Copernicus, Galileo's guiding lights, have been through him revived to us. And Galileo dared trust his eyes rather than holy books (against all scholastic teaching—how could you believe a sense that told you were *behind* your mirror, an obvious untruth? etc.).

We have seen how little time the revolution from old to new took in asserting itself; it is equally important to remark, obvious as it may be, that without the preceding millennial evolution this revolution would have been highly improbable, if not unthinkable.

Thus far I have spoken only of society and history and cultural patterns; this is our topic of interest, and it is fitting that examples be dramatic to be effective. Let us look now at the same things, with the cooler mind of the impassive scientific observer of the twentieth century. We set thus to understand things by making models of them (in our present acception), and soon discover that the working of a model—its ''physiology'' so to say—is independent of its actual implementation—its ''anatomy'': the same model may describe a society, a nervous system, a living organism, a communication network—names and figures may change and not the essential picture. A social system can be described, in this view, by two sets of mathematical laws: one for instantaneous action, the other for the (much slower) change of system structure as a consequence of past actions, as well as of external influence. The time is not yet mature for an exhaustive application of this method to the unfolding of history; some of the consequences and results it would have can be inferred, though, from experience gathered in other fields where the thing has been done. I shall try to condense a few here.

The operation, instant by instant, of the first set of laws—let me call them ''decision equations'', I have called them ''neuronic'' in a different context—corresponds to everyday life and to that part of the evolution of the system which is dynamic in character, such as cyclic changes of state that, if stopped, would die away. The second set—I call them ''learning'' or ''mnemonic equations''—changes instead constantly, at a much slower rate, the structure of the system. An effective comparison: our moving through the streets of Tokyo, as against the change of the topography of the town that occurs through the years.

When men increase in number, their actions diversify more and more, and something new and seemingly strange happens: the system acquires, in all known cases, a *hierarchical structure*; should this not happen, the system could not ''live''. The mathematics of it would be entirely out of place here, and is still in its early stages; to understand ''why'', though, is quite intuitive. Just think of a crowd and of an army: only the latter is capable of strategy and tactics, that is of ''intelligent'' action, because it has hierarchical levels. I have shown elsewhere with not much technicality that a simple and plausible criterion of economy accounts for not only the basic features of level formation, but also for the relative population of each level. The formation of new levels is what I call a ''revolution'', and is brought about by factors such as mere numerical growth, or by ''innovation'', that is by changes induced by spontaneous activity or external influence. The name in physics for this phenomenon is "phase transition": you may compare the slow heating of water with "evolution", its change, from ice to liquid to steam, with "revolution". I believe that the main reason for having hierarchical levels in a society or an army lies in the fact that a man can communicate efficiently only with a small number of other men, between, say, five and ten. But then the computer, taken as a symbiotic part of the new society, may destroy the very foundation of its present hierarchical structure! This may well be a major cause of the next revolution in our culture.

A last remark on levels: each must have its own logic, or *Weltanschaung*; the carpenter has nothing to learn from the nuclear scientist. It is *not* true that a few fundamental equations of basic physics could contain everything, if known: a TV set could *never* be read spontaneously out of Maxwell and Lorentz's equations. Finally, "forgetting" is as important as "learning"; we learn most new things by forgetting old ones, the extremely subtle skills of stone-age tool makers are forgotten, most of the mathematical skill of past century is lost except in books, yet we have become more powerful because we can operate at new higher levels, and this more than compensates for our loss of knowledge at lower levels; loss of memory and formation of new levels imply greater parcelisation of work and guarantee that the amount of individual work does not increase.

To conclude, I have only to ask the reader to compare these statements with the historical scenario which I have presented before: he will no doubt understand my plan, which was to present a well known situation, analyze some of its logic, and then use it as material to render intuitive things of which the mathematical discussion is not advisable here, or possible yet. If our concern must be with future society, then we must understand first the old; if it is change, we must know first about the major changes that already took place. What can we learn? Perhaps something: prejudice and irrationality and blind belief had to be swept away, and it took two thousand years plus Galileo. But here we live in a world where, on equally irrational grounds, there are countries that are ridiculously small and other monstrously large; contending beliefs, when all men need the same facts, irrespective of beliefs—I had better stop here. I want just to say, that we do not have to search far to identify the counterpart of Aristotelianism in our time: the way to tear it down is, rather than to fight it, to start planning a world in which harmony may find finally a rôle. To do so we must be aware of at least some of the laws that, by necessity of Nature, govern our society.

CHAPTER 6

A Model of Interdisciplinary Research in Order to Make Prediction

T. T. SEGERSTEDT

SCIENCE of the future must be about the expected conditions of man in society and the likely development of society. We do not claim to predict the future of individual man, but we believe it to be meaningful to predict possible futures of society if alternative conditions are created by scientific discoveries or applied technical results of research. According to our view, the future of man in society depends on those fields of research which are given priority.

If we accept the idea that the object of futurology is man in society, our next step must be to determine the nature of society. My definition implies that I regard society as a group in the sociological sense, but I admit that it is possible to describe society from other aspects as well. In my understanding of society as a group I mean *two or more interacting persons with common social customs, obeying social norms emanating from one and the same norm-source*. This definition implies that a group must be a system of communication, that is, a group is two or more persons in symbolic interaction and cooperation. When we study a group we can observe the activity of the norm-source, when pronouncing or when sanctioning social norms. Furthermore we can observe norms as linguistic phenomena and we can finally observe the uniform behavior and interaction of the members of the group, that is, the social customs of the group.

These elements form a system and are independent, and that is the reason why we talk about a social group. The group may be of various structures and have different functions but all groups must have a norm-source, common norms and uniform behavior. *Society is a group in this sense.*

There are different kinds of groups. Society may be called an inclusive group, because inside society there are several specialised groups with norm-sources, norm-systems and uniform behavior of their own. There are groups which do not belong to a single society. Such a group is the scientific community. It has got an international structure. We have a universal body of knowledge and a common language which makes it possible for a scholar to communicate with colleagues all over the world. The scientific or academic community can be described as a communication system, that is, a social group in the sense I have given the concept group. The unique quality of the academic community is that its norm-system always is international and the norm-source can, consequently, never be national rules. That is the true basis of academic freedom.

My ambition is not only to make a morphology of groups, I also want to introduce the dynamics of social reality into my theory. In order to do so I have to ask the question: what are the goals and basic functions of society as a group? In all groups we can point out three basic functions which must be fulfilled if the group is to survive. (1) reproduction, (2) socialisation and (3) production of goods. That does mean that new members must be recruited to the system and that that recruitment, in society, ultimately must be done by biological reproduction. Furthermore new members must be socialised, that is, introduced in the common system of symbols and learn the common social customs. Finally, in all groups a goal must be realized. In society that does imply that food, shelter and instruments of various kinds must be produced. In an agricultural society the three functions are carried out in the same group, the family, and such a society may be called a one-group society. In an early industrial society the productive function is separated from the family and carried out in a special work group as, for example, the factory. Such a society may be called a two-group society. In a modern post-industrial society we have got three groups as most of the socialising function is taking place in schools, colleges and universities. The post-industrial society may be characterised as a typical three-group society with the emphasis on the educational group. It is important to observe that there is a constant interaction and interdependence between the three basic functions and the groups surrounding the functions, that is, between the family, the educational group and the work group. This interaction may be illustrated by the following example: If a country is industrialised the consequence is that there comes into existence a particular work group. Because of this new group the family loses one important function, the productive function, and will for that reason be less stable and less autonomous, its norm-source has less power and there is less homogeneity in such a society. Industrialisation means furthermore that new sources of energy are being used, and because that means new instruments of production, you must be trained to use new instruments and new ways of communication. Industry creates new and wider markets and demands, and for that reason, people trained in foreign languages. All this implies that the family is gradually losing its socialising function and education is being taken over by schools, colleges and universities. So there will be specialised functions around the socialising function as well as around the productive one. That is the same as saying that if we start to industrialise and technify a society, the whole social structure will change: there will be a development from extended families to nuclear ones and the educational and productive groups will increase in importance. My belief is that the development may be said to be subjected to a kind of social natural laws. The laws may be expressed in the following way: if you change one of the basic functions of a group, it is necessary for the group to adjust all its functions to new situations in the biological and physical environment as well as to the internal organisation.

In a modern post-industrial society we may, as we have said, notice an increasing demand for skill and highly qualified persons. The increasing demand for skill means a growing importance of the educational group, which implies that more persons are employed in the educational group as teachers, technical assistants and administrators. The educational group is to an increasing extent, because of that, an occupational group for many people. It is also evident that more children are going to school and the time they are spending in school is increasing. By these illustrations I want to point out the inter-relations between the different functions and that interference with one function will be of consequence for other functions as well. I have pointed out that changes in the biological and physical environment of the group make it necessary to adjust society to new situations. In every definition of society you must underline that society is an organisation for human survival. The time dimension is always present in society. Although you may say that society is turned towards the future, the past is present in our ways of meeting the demands of expected events. The link between the past, the present and the future is education and scholarly endeavours. All social behavior may be said to be future-oriented and that is in particular true of scientific research. That creates obligations, as we may assume that our growing mass of knowledge to an increasing degree will determine the structure of our future.

It is evident that society is getting more and more dependent on research and scientific facts. That does mean that the borderline between basic research and applied research is getting very flexible. There will be a rising demand of highly qualified people to carry out applied research, as applied research must be done by first class scholars. I believe it to be a very dangerous misconception or prejudice to think that applied science is less demanding with regard to design, methods and theory than basic research. I should rather claim the opposite, as the result of applied research is going to be used in society and consequently will be of great importance for all citizens. On that account it is meaningful to ask the question: what is the essential difference between basic and applied science, that is, what is the difference between autonomous and nonautonomous science?

By autonomy I mean the right of scholars to decide and choose (1) the object of their research, (2) the methods to be employed in their research and (3) the right to estimate the value of the results obtained, that is, the value of the results from a scientific point of view. The value of the results for applications may be decided by others. But I think we must remember, however, that a result may be quite correct from a theoretical point of view but of little interest or value from a purely scholarly viewpoint, because it does not imply anything really new either with regard to methods or facts obtained. It may however be useful from practical or applied aspects in some planned survey. If a result on the other hand is dubious from a methodological point of view or not quite verified, it can never be said to be good enough for applied projects. I think we can state that two factors are common requirements of all scientific research, basic and applied: (1) a general theory and defined concept as a frame of reference (without comprehensive theory there cannot be any accumulation of knowledge, only scattered bits of information), (2) a generally accepted method and research technique which permits verifications of the derived hypotheses by all competent scientists in the field. No legal, administrative, political or economic authority could induce scholars to accept the results of research that does not qualify in these respects.

An important principle to which not enough attention has been paid in social science follows from this: the level of the theoretical system of science, in terms of inclusiveness, coherence, etc., determines the degree of autonomy of the science. No politician of today would try to influence a scientist working in the field of physics or chemistry inasmuch as these sciences have developed strong theoretical systems. But as long as only miniature theories or theories of middle range are available in social sciences, a considerable risk of exertion of outside pressure remains. Therefore, I find it dangerous that social scientists have so little concern about theoretical and epistemological problems. From this fact, I draw the following practical conclusions: the stronger the theoretical framework of a discipline, the less will be the autonomy risk for the scientist participating in applied research. In a context such as this, aspects of theoretical interest will be present in problem-oriented science or applied research, and highly theoretical research will be valued because of the significance of its results for applied research.

With regard to futurology, this means that the theory and the evaluation of results must be carried out by scholars according to the international norms of the scientific community. Politicians and administrators can never establish themselves as norm-sources of scientific futurology. With regard to the object of research, the situation is a bit more complicated. A national government has got the right to decide if it will give priority to atomic research or cancer research and make funds available for such purposes. But with regard to futurology, there is no sense in discussing the future of Sweden or Japan apart from the future of the rest of the world. I think we have to admit that the future of mankind is dependent on the science policy of the superpowers. That is, there will be another future for all of us if they give priority to atomic research or if they give priority to research in life sciences. I believe it to be the duty of scientists to make that clear. Consequently: although the aim or goal may be said to be a practical and applied one, the scientists must regard it as their duty to choose and determine on an international level the objects of future research. For that reason it may be important to discuss the relation between research and the future. It may be fruitful to talk about different levels or strata in our knowledge about the future: (1) the first and basic stratum is the one described by established laws of nature. We know for certain that the sun will rise next morning, that water will boil at the same temperature next year as today, that electric currents will behave in the same manner as before, that there will be a total eclipse of the sun on 11 August 1999 and so on. (2) The second level or stratum is formed by human decisions in the past. The decision 10 years ago to exploit certain natural resources will influence life for generations. The same is true with regard to our educational or agricultural policy. (3) The third stratum consists of our decisions of today and their consequences. If we really could control all the results of our actions of today and predict their consequences, the importance of stratum (2) would gradually diminish. With regard to stratum (1), our possibility of making predictions is dependent on our knowledge of the laws of nature. But it is really strata (2) and (3) which are creating the problems of forecasting and predicting. When we are discussing our possibilities of making predictions we often say that we have traces or signs of the past, but that there is no sense in saying that there are signs or traces of the future. But is that really a correct statement? We certainly have signs and traces of the past. Such signs are buildings, monuments, inscriptions and documents. But I think that we may regard buds on the trees, the roots in the soil, pregnant animals, as signs and traces of future events. They may be described as elements of the future, but they are not causes (or they are not the only causes) of the future. We would perhaps make the following statement: all conditions being equal, these trees will, because of the buds, be green next spring. On the other hand it is possible for us, if we have sufficient knowledge, to manipulate the buds or the roots and in that way change the future. It is evident that man, because of his basic needs, has changed his natural and social reality and he can, if he has got sufficient knowledge, manipulate the roots, seeds, animals and his own biological dispositions. He is in that sense communicating with the future and structuring it. He sends messages about his present hopes and demands and how he hopes to get them realized and satisfied in the future. There is, furthermore, a feedback on his present situation. His expectations or predictions of future events influences his present decisions. There are two elements determining his decisions and actions: the knowledge of his present situation and the expectations of the future. His knowledge of the present situation and of the future is determined by his scientific progress and endeavour, that is, by his knowledge about the true relations between phenomena. It is important to understand that the future is not just something passively expecting us; it is partly created here and now by our actions.

The problem with regard to our predictions of the future and our adjustment to future life is the difficulties of coordinating and surveying our rapidly increasing knowledge. For that reason it is necessary to work out and develop a theoretical system by which we can formulate hypotheses about future consequences of scientific discoveries with regard to our physical and social reality. If we do not have such a general frame of reference, research may just be one instrument for creating social and environmental catastrophies. The frame of reference must be international, since that is the only way of influencing decision-makers of the scientific superpowers in their setting up a priority list of different fields of research. I have already pointed out that it is the science policy of the superpowers which will determine the future of mankind. For that reason it is an urgent duty of the international scientific community to create a theoretical frame of reference which makes it possible to make predictions of the most likely results of priorities given to different fields of science.

My suggestion is, that we can use my concept of society as a starting point when trying to build a model of interdisciplinary research in order to make predictions. I have already said that there is an interaction between the three basic functions and between the three groups in which the functions are embedded. I have pointed out, for example, that if we change the structure of the work group because we have introduced new sources of energy, that will probably also change our expectations with regard to the functions of the educational group. That may in its turn be of consequence for the family group. Work may demand for instance, more geographical mobility, which will influence family life. There are other illustrations of my thesis; science has made it possible to control the birth rate. That means that the size of the family is, or may be, reduced. The consequences will probably be smaller classes in school or an overproduction of teachers. We may have to face new demands with regard to housing and there may be a lack of labor force, and because of that factories must be still more mechanised. We know for certain that in the post-industrial society there are increasing numbers of people employed in service work and less people in direct heavy production, which will be of importance for our educational policy. But this knowledge has never been systematically analysed from different aspects of knowledge; that is, it has never been interdisciplinarily studied from a common frame of reference. Such an interdisciplinary approach is to ask the question: what does, for example, the possibility of manipulating the genetic code mean with regard to family life? If we can improve the technique of transplantation, what would that imply with regard to the average age of man in the future? What does the manipulation of the genetic code and new knowledge with regard to transplantation mean for medical care etc.?

We must try to find out not only the biological consequences of such scientific innovations, but the social implications as well. If we make futher progress with regard to the electronic science we have to ask: what does that mean with regard to communication, with regard to social control, to international understanding? How does it influence our economy, our social policy, our political life and our way of educating?

We know for certain that we can expect a lot of new discoveries in life science, in physics, in chemistry and physiology. For that reason I believe that we ought to ask prominent scholars, as has already been done by Rand Corporation (U.S.A.), what new discoveries they expect in their fields of research during the two, three or four coming decades. After that has been done we ought to try to find out what such breakthroughs would mean with regard to the basic social functions and their groups.

How would a breakthrough in brain research influence education? Does such an influence on education mean anything for the productive function and could that be of importance for family life? In such a way we could discover alternative futures and we could have a chance of making priorities with regard to our valuations.

I call such an approach *scientific strategy.* We know that military staffs are playing what they call war games, that is they try to plan their future actions with regard to possible alternative war situations. In the same way scholars should try to predict what is going to happen with the basic functions and their groups and consequently with society as a whole if priorities are given either to life sciences, or to atomic physics, or to electronics, or to some other field of research. But such a scientific strategical game can only be played if representatives from various academic disciplines are taking part. That is why interdisciplinary research inside a common frame of reference is so necessary.

CHAPTER 7

From Exchange to Communication: The Birth of Programmed Society

A. TOURAINE

Introduction

Industrial Society is in a state of crisis. Cyclical crises previously seemed to be an integral part of the capitalist mode of industrialisation; alternate phases of expansion and recession, or the longer phases of Kondratief's cycle, structured the history of industrial economy. The state of crisis was thus seen both as a disease and as a mechanism of growth. In any case, it was a part of a particular type of economy and society: those who believed that the capitalist system would degenerate into a general crisis and would disintegrate as a result of its internal contradictions, after being apparently justified by the great crisis of the thirties, saw their predictions disproved by 20 years of exceptional and virtually uninterrupted growth.

Nowadays, when we talk of crisis, what we are referring to is of an entirely different nature, or rather it includes two quite distinct kinds of phenomena, neither of which can be accounted for merely in terms of the degeneration of an economic system. In the first place, it is the worldwide leadership of the capitalist nations that is now at stake; on the one hand as a result of the sharp increase in the price of various raw materials from developing countries; on the other due to the massive export of capital and factory plant from the high-wage to low-wage nations eager to encourage new industries. In the second place, a whole system of values is being challenged. This crisis of civilisation was in fact becoming apparent several years before the critical recent events in the economic world had taken place, namely the 1971 dollar crisis and the steep rise in oil prices following the Yom Kippur war of 1973. The meetings and declarations of the Club of Rome, Ivan Illich's books, and the student movements in Japan and the United States, then Germany and lastly France and Italy, were the most spectacular manifestations of this crisis of values, which, as generally happens when one particular type of culture and society gives way to others, gave rise to various utopian concepts. By utopian concept I mean the concept of a necessary and desirable order rather than an order which has no justification beyond individual interests, political compromises or the constraints of force. An order which is necessary because of the natural limits of growth, the thresholds beyond which progress becomes repressive, and the laws of stability of the ecosystem which no species may transgress without incurring the risk of extinction; an order which is desirable because it substitutes basic needs, voluntary associations and communal relationships for the false logic of bureaucracy, the ravages of technical progress and gigantism, the horrors of totalitarianism. More suggestively, we sense that the nature of authority and the methods of socialisation, attitudes towards work, and concepts of the future have undergone radical changes. The crisis affects not only institutions, but equally our motivations and our social behaviour. It is indeed a crisis of civilisation as a whole, not one affecting merely the functioning of a particular area of social organisation.

What is the nature of this crisis? We are moving away from the outskirts of industrial society, but whither? Are we just to drift, perhaps to land on a Cytherean island, or perhaps to enter a hyper-industrial society? Are we living through a period of decadence, or are we witnessing the tail end of several centuries of growth and the return to societies essentially preoccupied with their own stability, as were our rural societies not so long ago in both Asia and Europe? Or are we perhaps moving towards a society marked by a greater capacity for self-modification than Industrial Society, that might provisionally be called post-industrial society? There is no doubt that the threat of decadence exists. Accustomed as they are to plenty, our societies are tired and irritable, preoccupied with self-protection and enjoyment, and perhaps allowing themselves to be dragged into a degenerate future as new Lower Empires or a latter-day Byzantium.

This is surely only one of the possible answers to the challenge of history; not only are there others, but the study of these collective movements does not preclude our asking what problems they do, or do not, resolve. The great choice remains: are we facing a new phase of growth or a return to equilibrium, to a post-industrial or indeed a post-historic society; are we at the end of a development phase, or the end of development itself? We have just lived through a vast counter-cultural movement, which has directly questioned the values of industrialisation and growth, and which has sought stability and identity in any form. I believe this could only be a brief transitional phase between rejection of the values of industrial society and awareness of the transformation of the whole economic order. We must understand that the "exhaustion" of industrial society, in the countries where it had achieved its greatest development, may herald the difficult but indispensable transition to a new type of society, more active and mobile, but also more voluntaristic and dangerous than the society we are leaving behind us.

Levels of Historicity

A self-confident society considers itself to be the ultimate product of history, and only defines other societies by the lesser degree to which they possess an attribute identified with historical progress or with the essence of social existence. Merchant societies contrasted civilisation with barbarity and prided themselves in being the most civilised because they had more formally defined the rights governing people and possessions. Industrial societies have more recently explained their superiority through the greater number of their mechanical slaves, the available power of their machines. In the same way, we are already beginning to class societies, or at least countries, according to their capacity to process information, insofar as this capacity is indicated by the number or cost of their operational computers—an indication that we have already entered post-industrial society, despite the fact that the vast majority of the population is still employed in an industrial or market economy. Such an approach has above all the merit of indicating the shifting of human activity, from one type of society to another: from commerce to manufacture; from manufacture to communication; of introducing the idea of passage from one type of culture and society to other types; from one type of relationship between a collectivity and its environment; and one type of relationship between social actors, to others. This disproves the previously and generally accepted idea of evolution in terms of increasing differentiation, enrichment or rationalisation. The rift separating trade and industry, which has been analysed from many different angles, reappears between industrial and post-industrial society, to use—however provisionally—a superficially historical definition. But each of us resists the idea of a succession of entirely different societies, especially when the transition from one to another has taken place smoothly. We can easily accept that a conqueror could transform a society, and that the closure of trade routes, invasions, and the decadence of the Roman Empire could have brought about the enclosure of vast domains and the birth of feudalism; but Great Britain, the Netherlands and France, to mention but a few, developed without so sudden a transition from merchant societies into industrial societies, and almost everywhere there were industries already extant within merchant societies. Neither do the more ardent supporters of the notion of a post-industrial society deny that it could be considered, at least partially, as a hyper-industrial society. How then can we combine discontinuity and continuity in the succession of societal types? It is necessary to answer this general question in order to situate the communication society that is forming before our very eyes in relation to the manufacturing society in the midst of which it is appearing.

A society acts upon itself in a primarily material way. It sets aside part of the available production and re-invests it in the productive system. True, there are some societies which are not self-modifying in this way, but these might be called societies without historicity—which does not mean without history, if we define historicity as action which prevents reproduction of the established order and therefore demands a capacity for transformation and development. These societies are ruled by models which prescribe the equilibrium and control of the exchanges within the social order, or with its environment.

Beyond such societies, which may thus be defined by their structure, we can distinguish a succession of historical types, that is to say societies defined by a certain level or mode of historicity.

Many societies are self-modifying only at consumer level, that is to say they invest in order to increase the quantity of available products by increasing the available workforce or animal power, and by extending cultivated areas and increasing their yield. These may be called Agrarian Societies. They have little historicity, for they undergo self-modification only at the final stage of their economic system, which may be simply described as a co-relation of production with collective organisation of the means of production with a "social" mode of distribution, and with individual or collective consumption patterns. Historicity in agrarian societies is therefore only found at consumer level.

Merchant societies, on the other hand, undergo self-modification at distribution level. By organising trade and transport of goods, guaranteeing the quality of goods supplied and the value of currency used, by maintaining order over trading routes by land and sea, and by formalising legal rules, merchant societies also allow specialisation of production, leading to greater efficiency (and profitability). However, manufacturing itself is not directly affected by commercial progress. Craftsmen working at home or in workshops continue to respect professional traditions and to use their skills in conformity with the rules of a corporate working organisation. This clearly shows the limits of self-modification within a merchant society.

If we talk of industrial revolution, however unsatisfactory this expression may be from the historical point of view (at least in the case of European countries) it is because the transition from trade to industry indicates that historicity has spread beyond the fields of consumption and distribution and penetrated the world of production. But we must again note that industrial society does not invest directly in the production of production, and is not therefore directly associated with what has been more recently called a scientific and technological revolution. By definition, industrial society represents revolution, through investment, in the *organisation of labour*: this is the process described by Ure and Marx in their celebrated writings. The Industrial Revolution begins the moment the merchant no longer merely controls the product of the craftsman's craft, but also organises the workers' working system (factory) and the conditions of collective manufacture, by imposing new methods of work division and specialisation. Industrialisation is not defined by the use of machines, but by the organisation of labour. It is not by chance that one of the first philosophers of industrialisation, Saint-Simon, gave the name *"l'Organisateur"* to the journal he published during the early years of the Restoration; nor is it by chance that towards the end of the century the phrase *organisation of labour* was used to describe the activities of the two great figures of mass industrialisation, Taylor and Ford. Rational or scientific organisation of labour (as it is sometimes called) is not a mere characteristic of industrialisation, it is the very definition of it. This means that industrial society, like any other society, cannot be defined in terms of technology, productivity level, or income per inhabitant, but rather by its mode of self-modification.

This definition of industrial society calls for the concept of a post-industrial society which would promote historicity, and primarily investment, to a level not attained by industrial society, that of the production of production. The organisation of labour only affected the relationships of the workers among themselves, and therefore the level at which manufacture took place. Post-industrial society acts more globally at managerial level, i.e. on the overall mechanism of production. This action takes two main forms. Firstly, it is innovation, that is to say the capacity to produce new products, in particular as a result of scientific and technological investment; secondly, it is management itself, that is to say the capacity to make use of complex information and communication systems.

Obviously, these levels of historicity do not form a simple succession; on reaching a certain level, a society continues to exhibit lower levels of historicity characteristic of preceding societies. An industrial society does not forego the benefits acquired through trade, and a post-industrial society does not forego the organisation of its labour force; but it is in the nature of a particular society that it recognises the greater importance of a certain type of investment and chooses in its favour whenever it conflicts with some other type, characteristic of some other type of society. Thus a post-industrial society cannot be defined, any more than an industrial society could be, by a particular technology. It is just as superficial to speak of a "computer society" or a "plutonium society" as it is to speak of a "steam-engine society" or an "electric-motor society". There is no reason why so great a privilege should be granted to a particular technology, whatever its historical importance. The important point is to recognise that a post-industrial society is one where all the elements of the economic system are affected by the actions of the society on itself. These actions do not usually take the form of conscious will embodied in one person or even in a group of people. This is why such a society must be termed a *programmed society*, a designation which fully indicates its capacity to create models for the management of production, organisation, distribution and consumption; thus a society of this sort appears, at all levels of operation, not as the result of natural laws or of specific cultural characteristics, but rather as the result of the production, by society acting upon itself, of its own systems of social action.

Programmed Society in Human Experience

We must now change our perspective to the viewpoint of those who are living in this society and who, particularly as individuals, behave towards it as consumers rather than as producers. This is where quantitative analyses can help us—in the same manner that, in the course of industrialisation (what Carl Polanyi called "The Great Transformation") it was possible for sociologists to consider not the organisation of labour, but rather the differentiation and growing rationalisation of social relationships, and still more simply, increased exchanges within a social unit. This is what Karl Deutsch in the US and Gino Germani in Buenos Aires recently and almost simultaneously termed a society's *degree of mobilisation*. A programmed society involves a noticeably higher degree of mobilisation than an industrial society. In the latter, individuals were drawn into directed systems of collective organisation almost solely in the field of employment, though sometimes also—but in a much more limited manner—in connection with housing. It is characteristic of a post-industrial society that it sets up large centralised management systems in the most diverse areas of social life. There has been talk of the industrialisation of information, of consumption, of health, of scientific research, even of general education. In fact the word is inexact, but it is true that decision and management centres are being formed which are capable of producing not only systems of "means", but also goals for social action; systems of care and treatment are presented as models of consumption or as educational curricula.

This extraordinary acceleration and multiplication of more or less intensely programmed communications has already given rise to some very positive attitudes, that should not be eclipsed by the force of counter-cultural protest of recent years. In the first place, the majority of participants in this type of society are attracted by the multiplication of information and the consequent multiplication of available choices. There is no justification here for citing in opposition the concentration of decision-making and the uniformisation of programmes or messages : where such a uniformity does exist, it is not due to the intrinsic nature of a programmed society, but (which is entirely different) to the nature of the political and ideological control system which forms part of certain national societies. We should beware of facile and merely elitist criticisms of the mass-media. Those who in the past enjoyed a great freedom of choice with regard to material or intellectual consumption now enjoy a still greater freedom of choice—as witness the intense interest in access to cultures temporally and spatially distant from our own. Those who were imprisoned within the confines of parochial existence, subject to the influence of the traditional educated classes and the limits of popular literature, now have access to a much wider and more open repertory. Programmed societies cause people, goods and ideas to circulate to a far greater extent than did the preceding societies.

Again, we must recognise that societies possessing a high level of historicity, and a high capacity for self-production, considerably reduce their capacity for self-reproduction. It is even paradoxical to find so many critics insisting so much on the importance of the mechanisms ensuring reproduction and the maintenance of the established order in our type of society. It is my contention that this is an indirect and almost paradoxical expression of the growing awareness of mobilisation in our production societies. These societies are increasingly hostile to anything that seems to be oriented towards reproduction, despite the fact that less and less conduct, ideas and behaviour patterns are being reproduced. This is often brought to our notice in the so-called liberation movements. Whether simply liberal, libertarian or revolutionary, their principal direction seems to be towards the destruction of acquired positions, transmitted roles, and "ascribed" statutes, thus allowing all categories of the population to participate in more and more intense exchanges and communications. The most noteworthy of these is without doubt the feminist movement. While it is true that one may see in it other meanings (which to my mind would be even more important if they were more militant in nature, and challenged particular aspects of social power) the most significant aspect of the feminist movement is undoubtedly the fact that it accompanies the general crisis in reproduction systems.

The most visible aspect of the feminist movement is thus analogous to the movement instigated by the prophets of capitalist industrial society against slavery and against anything which might oppose the free circulation of people on the labour market. Here lies the ambiguity of a movement of this sort. It does indeed break down barriers, prejudices and taboos, but we should not allow ourselves to be misled by the revolutionary or libertarian language it often employs. It is already noticeable that the breakdown of barriers between private and public life, the "publicity" given to sexual conduct and the more egalitarian participation of women in the economic and professional world, can be largely explained in terms of the interests of a consumer society which needs to increase the available income per household and to expand the areas of behaviour that can be influenced by centrally established models.

This ambiguity merely points to the well-known fact that cultural innovation is at first closely linked with the formation of new leadership groups. On the other hand, dominated social categories remain culturally defensive, embracing anything that enhances their particular identity and appears to offer protection against (new) external domination. Cultural progressivism can be associated with social conservatism; this is why groups of women wishing to oppose this conservatism are obliged, in order to maintain their image as cultural innovators, to dissociate themselves in a particularly spectacular or violent manner from the ruling classes and their behavioural patterns.

It is nevertheless true that the behaviour patterns just mentioned are part of a modernising attitude, tending towards a more open society and therefore constituting a new manifestation of the philosophy of enlightenment. On the other hand, both in the non-ruling social categories and in broad areas of the intellectual sphere—especially among those whose level of education is above the "economic" level—the fear is often expressed that individuals and groups will be imprisoned in ever more complex networks of signals, rules and prohibitions. Expressions such as "meshed" or "cable-linked" society, are heard in references to increasing pressures towards normalisation. In pre-industrial and culturally traditional societies, certain rules were maintained or taught in a very authoritarian manner, but they were part of a loose network and there were many areas of behavioural indetermination. In a modern city it is literally impossible to take a step without receiving orders, without being exposed to publicity or propaganda, without being confronted by scales and systems for self-evaluation of social level. This is why there is such a strong desire for non-social inter-personal relationships, or to build communities representing so many "holes" in the ever more closely-knit social network. Thus marginality, considered for a long time as a failure to integrate, becomes a sign of opposition, a laboratory where a new culture, a counter-project of society, could be born.

At another, more global level of criticism, the idea appears that a programmed society increases the distance between the leaders and the led. It is true that the idea is repeatedly expressed that the new technologies allow greater decentralisation. This same idea was expressed towards the end of the nineteenth century when many social philosophers contended that electricity would encourage decentralisation just as coal had meant centralisation. In fact, the idea that technology rules society is yet again found to be wrong. It is not in the nature of electricity to determine how it is used in society; and the same is true of the computer. On the other hand, it is true that the creation and increasing diversification of production and information management mechanisms leads to a concentration of power. Such concentration has long been a highly developed feature of industrial organisation; it continues to progress, but above all it is the power to make decisions in other areas of social life that has undergone ever greater concentration. A frequently studied example is that of scientific research and the transition from "little science" to "big science", especially following the fearsome arms race between the two superpowers. It could even be said that a programmed society permits and encourages a greater interdependence between the mechanisms of domination, apparently justifying de Tocqueville's classic idea in the second part of his American journey. Modernisation, the destruction of roots and the increasing pace of change, do these not lead to the overwhelming authority of a central power capable of imposing its will on a society fragmented by the destruction of communities and traditions?

Beyond these positive or negative ideas, the most important thing is to tackle the question raised earlier of the nature of the relationships between technology and society. Have we become a "technologised" society? Can we expect technical progress to lead to social progress by increasing living standards and multiplying the number of available choices? Or should we, on the contrary, recognise that certain techniques have become a direct menace? Questions of this sort are of course unprecedented, and the definition we have given of a programmed society no longer allows us simply to draw traditional conclusions, as handed down from the 19th century, concerning the effects of technology. It is no longer possible to say that technology and science have positive effects of themselves, but that these may be turned in a negative direction by the social use that is made of them. If this were the case, it would be sufficient to exchange that use for another—to replace oligarchy by democracy in order to transform death forces into life forces and sources of power into fountains of well-being. Such a distinction between the productive forces and the social relations of production is nowadays devoid of any meaning. It could only apply to an industrial society in which social power had not yet begun to affect the level of production itself, but merely that of labour organisation. But today the choices are more radical, and it is in the nature of the ruling powers to be identified with the development of the productive forces, the management of information systems. However, this simple observation in no way implies that our society is directly dependent on the technologies it uses.

The time has in fact come to reverse the traditional logic. When anti-nuclear campaigners state that the nuclear industry is leading to a centralised and authoritarian society and that a plutonium-based civilisation necessarily implies a police state, whereas a solar society would be democratic and decentralised, their logic is standing on its head. In fact it is the power wielded by the main decision-making bodies that imposes a particular type of energy policy. In France, it was the powerful position of Electricité de France and the Commissariat à l'Energie Atomique which led to the adoption, in 1973–74 and in the absence of any truly democratic debate, of an all-nuclear policy. This policy was in fact a logical development from the all-electric policy imposed some years previously by EdF, for commercial reasons. It is in this way that technological choices are also political choices, and their effects are merely indicative of a relationship between the social forces involved. This is not entirely novel: for over thirty years industrial psycho-sociology has emphasised the fact that the effects of technical changes are less dependent on the necessary consequences of these changes, which are generally very difficult to determine, than on the social mode of their introduction. Studies of the changes introduced in administrative work by new methods of data-processing show even more directly the impossibility of identifying the specific consequences of these technical changes. The extreme diversity of the reactions, and the wide variety of changed situations observed, show that it is no longer possible to isolate a technological cause in a programmed society, in which it is characteristic that social power operates directly at production level, at the level of models for the determination of both demand and supply. It is precisely in the case of societies most dissimilar from programmed society, even from industrial society, that we can speak of technological determinism inasmuch as the society's process of self-modification had little effect on technology and production, and a much greater effect, as we have already noted, on consumption and distribution. We can quite reasonably speak, as Roger Bastide has in the case of Brazil, of a civilisation based on wood, or of others based on coffee, cocoa etc. On the other hand, in industrial society, we are witnessing the disappearance of occupational autonomy, of specific cultures linked with particular trades or products. Occupational activity is increasingly defined in terms of the place occupied in a system of communications, and psychologists are already measuring levels of skill in these new terms. Again, we must add that the structure of this communications system is not itself determined by the technology involved, but is increasingly determined by the socio-political relationships within the organisation.

This leads to two more general observations. Firstly, the advent of programmed society is a new and decisive step towards the decline of society as presently known. Are we not witnessing the emergence of an increasingly restricted notion of society, both as an analytically defined whole and as a particular geographical and historical reality? Pre-industrial societies spoke in broad terms of Roman or Athenian civilisation, which clearly indicated the direct link between a societal type and a political unity. When we talk of industrial society, we are aware of a considerably greater distance between it and a national state; nevertheless, we still use the same word in speaking of industrial society or, for example, of French society. When talking of programmed or post-industrial society, it would be more correct to replace the word "society" by "culture" thus denoting the increasing dissociation between the management of a collectivity's relationships with its environment, and the determination of social order. We shall return later to this central problem; let it suffice here to say that the increasing complexity of a programmed society means that its degree of integration can only decrease. A programmed society corresponds to a much more complex, less mechanical and less stable model of organisation than pre-industrial societies. We can all sense this. Different experiences of participation in this society cannot be traced back to the same central point, but rather to centres of decision situated at very variable distances from this theoretical point; in fact, these centres of decision more probably constitute a system without a central point, in contrast to the "pyramid structure" of societies having the lowest levels of self-modification of their functioning.

The second general conclusion we can draw from this analysis is that the idea of ordered society, cherished in the past as was order itself in relation to chaos, nowadays appears to be fraught with danger. From the moment when social power takes the form of intervention at all levels of collective existence, it is bound to seek the universal rather than the particular, as happened during the French Revolution. We are therefore facing an extremely rapid reversal of attitudes, leading to an increasingly positive appreciation of individuality, difference and uniqueness, whilst order appears synonymous with imitation, uniformity, loss of information and therefore, ultimately, with disorder and entropy.

New Class Relations

We must now return to an expression used at the beginning of this analysis, one that is central to our argument. What is meant by society's process of self-modification? What is this society that is thus considered as a person, even as a god? In fact, it is difficult to imagine a society or social whole undergoing a process of self-modification. This is the most constant object of a form of utopia that reappears regularly in the course of every century, and that might be called the populist utopia. It is hoped that the same can become something different—that change can take place without loss of identity. Albert Hirschman has analysed this very well in relation to social and political change in Brazil. It is in this sense that the nineteenth century Russian populist movement hoped to avoid the great social and cultural rupture caused by the capitalist mode of industrialisation, together with the proletarianisation it entailed. They wanted to use certain forms of traditional Russian organisation for the creation of a future society: pre-capitalist social patterns would have been transformed directly into those of an industrialised post-capitalist society. This desire to avoid rupture is fundamental, but it is as impossible for structural change to take place in a continuum as it is for a revolution to be accomplished by existing laws. Historicity, and the more material aspect of it which is investment, cannot be set in motion by a whole society. In fact, for a whole society to exist in the first place, it is necessary to construct systems for the maintenance of order, for socialisation, for reproduction, for social control and for repression. It is because these instruments of social order and maintenance of social order exist that historicity can only be introduced by a portion of society, a portion that is able to escape the constraints of the existing order or, as is more often the case, to use them to its own ends by redirecting them from their original purpose. The real controlling class is the group that takes charge of historicity, the particular "actor" which exerts the most widespread action on the operation and transformation of society. This controlling class, in identifying itself with historicity, simultaneously identifies historicity with its own self-interest. On the one hand, it is progressive, in that it brings into play a greater degree of a society's capacity for self-modification, and resists earlier dominant social categories and earlier instruments of social control. On the other hand it raises barriers to protect its own privileges, that is to say its identification with the interests of society as a whole. Each particular society possesses a specific class relationship. No single social actor, whether dominant or dominated, always plays the same part in class relationships, nor remains the permanent trustee of an eternal mission. The further we go from industrial societies, the more we find the distance between classes matches that between the internal agents of reproduction of the existing state and the external agents of change. The controlling class is also a class of warriors, both conquerors and defenders of the collectivity. At the same time, this social distance is institutionalised in the form of status "rules", transmitted from one generation to the next, which appear as strong segregations, prohibitions that are impossible to overcome, and innumerable signs indicating membership in particular social levels. In merchant societies, class relationships become more "civil", separating the productive craftsmen from the merchants who control the exchange system. The site of these class relationships is the labour market, because the craftsman must sell his products or his labour to the merchant: thus class relationships are for the most part shaped in the market place, i.e. in the town and within its legal and political organisation.

Class conflict in industrial society is between the worker and the organiser, that is to say the boss, and the scene of their conflict is the factory, or to be more precise, the workshop or workpost, in other words, the place where organisation imposes its law on the worker's trade. The vital breakthrough in the development of a working class consciousness occurs precisely when patterns of rationalisation enter large-scale industries, a process which threatens but has not yet broken the resistance of skilled workers, many of whom have become toolmakers.

What then is the nature of class relationships in a programmed society? At first sight we are tempted to say that the central conflict is between directors and employees, between those who design, formulate and direct production programmes, and those who execute or "undergo" them. We are witnessing a process of proletarianisation of junior and even middle-level white-collar employees, just as we saw that of the workers; new techniques will next threaten the professional status of executives, and even doctors and teachers. But in reality the division between those responsible for conception, and executive personnel who bring the concepts to realisation, merely delineates a stratification scale and therefore relationships based on authority.

We mean much more than this when we talk of class relationships. The dominant controlling class is that which has the power to guide the creation of cultural models and social norms; the dominated class is that which can only attain historicity in a subordinate manner, by accepting to play the part assigned to it by the controlling class, or else by attempting to destroy the monopoly of historicity held by that class. If it is the controlling class in programmed society which has the capacity to create models of social consumption, the dominated class cannot comprise those who execute and put these models into practice, but rather those who have to conform to them. Even if it is dangerous to say so, we must, in order to emphasise the separation between programmed and industrial society, recognise that the main social conflict of today is between the machinery of production and management, and the consumer himself. The conflicts of authority which exist within large organisations between directors and executive staff are by no means the most important. This is why the first manifestations of new social conflicts have brought consumers to the forefront in a spectacular manner, or at least have made them involved and concerned. Those who have spoken out against schools and universities in the name of education, against the politico-scientific machinery in the name of public welfare, against hospitals in the name of health, against urbanisation schemes in the name of interpersonal relationships, and against the nuclear industry in the name of ecology, have always championed what at first may seem like a form of consumerism against the influence exerted by the "big machine" on the determination of demand. The power of the machine is generally referred to as technocracy. Just as, in an industrial society, the controlling class is the *organising class*, whether the organisation is private or public, we must not confuse technocracy with state-dominated administration. There is both a private technocracy and a public technocracy, just as there is a capitalist and a collectivist technocracy. The strength of technocracy, as pointed out by Galbraith, is that it is a machine, not simply a willpower, and that it protects the interests of a technostructure rather than those of the owners of capital. Confronted by technocracy, the consumer speaks out in terms of his real needs, in contrast to the traditional logic of industrial societies which conceived these needs as being a simple copy of economic growth. Engel's famous principles attempted to show that an increase in income led to an increase in consumption of non-essential goods, and to a relative decrease in consumption of basic goods, e.g. food. In fact, a violent rejection of this quantitative concept of needs is now taking place. This rejection appeals to deep, fundamental, natural needs, notions which do not have clear sociological meaning, but which indicate the desire to counter technocratic demand modelling with another life-style, other preferences, determined by the will to regain control of the instruments of production and organisation in social life.

In a programmed society, social conflict is characterised by the fact that the controlling class appears to rule over all aspects of social existence, thus preventing the dominated classes from speaking or acting from the standpoint of their own social and cultural autonomy. These classes are therefore obliged to oppose social domination in the name of the only thing that could escape the grasp of the controlling class, namely nature. This explains the importance of the ecological movement, champion of life in the face of genetic manipulation, pollution and the dangers of nuclear contamination. It also explains the importance of protest movements based on biological rather than social status: feminism, youth (also old age), the fact of belonging to an ethnic group or even, to some degree, belonging to a local or regional culture (for example, with regard to language, since the same language can be used by a collectivity at extremely different levels of social and economic organisation, it cannot be considered as one simple social product). However, this protest action can become wholly utopian, resulting in total rejection of the concept of social life, if it fails to establish links with some counter-offensive movement, i.e. a movement endeavouring to channel modern science and technology into a counter-model social and political organisation. Such a counter-model can no longer restrict its impact to the field of work organisation, as was the case in industrial society. And since the idea of management has superseded that of organisation, it is natural that the concept of self-management should replace that of socialism, that is to say the concept of workers' control of the organisation of work. But the protest (defensive) action must link itself with the counter-offensive action by finding common ground between the two. In merchant societies this common ground of protest was called freedom, because it was concerned both with defence against the political and legal power of the merchants and with the creation of another order, equally defined in legal terms. In the industrial era, the common ground was called justice, being concerned with returning to the workers the fruits of their labour and of industrialisation. In programmed society, the common ground behind both protest and counter-offensive is happiness, that is to say an overall concept of the organisation of social life based on the estimated needs in the individuals and groups within the society. This means that the field of social strife is no longer as well defined in programmed society as it was in the preceding societies: in agrarian societies, everything was clearly related to the soil; in merchant societies it was the citizen, the inhabitant, who was the principal "actor"; in industrial society it was the worker. But in programmed society, it is the "social actor", in any of his many roles, so that one could almost say it is the living being. This is why protest movements act in the name of a collective entity, whether it is the individual considered in terms of his bodily nature and his projects, or whether it is the community itself. However, all the different facets that apparently invest social conflicts in programmed society with an exceptional vitality and widespread nature are also a cause of their weakness, for the generalised nature of these conflicts has incidentally deprived them of a common ground. To use a crude analogy, fires may start anywhere, but society seems less threatened than previously by a huge conflagration. This is perhaps the reason why the future development of social movements and conflicts is more dependent in this society than in others on the state of the political order, that is to say either on the intervention of political parties or on a crisis involving the state as a whole.

Society Without State, or State Without Society

Is this vision of programmed society too conservative? Ten years ago, when industrial society seemed to be at the height of its strength, it seemed rash to imagine that it might already be in the process of being superseded by another societal type. Today, the notion of post-industrial society has often been abandoned because it was apparently linked with the optimism of the industrial period and therefore reduced to the image of a hyper-industrial society, which was not a solid foundation for the idea of societal change. Today, awareness of a crisis of values and of industrial organisation leads one to make the opposite criticism of the idea of a programmed society, namely that this idea has been left behind by transformations that have already appeared. However, criticisms diverge at this point, because the situations already described may be subjected to two different analytical approaches, which will now be described.

As already noted, for some it is not a stage of growth which is reaching its final development, but growth itself, the very essence of development. The very great capacity which programmed societies have for self-modification precludes their pursuit of the destructive creativity which was the hallmark of industrial society. For a long time we had a finite capacity to act on an apparently infinite environment; today we are in a contrary situation—our capacity for action appears to exceed the resources on which it can draw. Should we not therefore think in terms of survival and stability rather than progress? Is it not time we realised that man is not in confrontation with nature, but that he is a part of it? The most extreme manifestation of this attitude calls for the return to a society of exchanges, a society without historicity. Certain anthropologists such as Marshall Sahlins have even spoken of a return to plenty, pointing out that our industrialised societies are dependent on scarcity and that the societies of the great North American plains knew how to maintain social stability by consuming or ceremonially destroying the available surplus.

Others, such as Pierre Clastres, have called for a return to stateless societies. On all sides there is a widespread nostalgic desire not for order but for an end to growth, which would give back some of their lost importance to personal experience, direct exchange, the life of the body and local organisation. Concurrently, the mainstream of protest, which seemed to be focused on social relationships based on power or ownership, now seems to be focusing anew on the social actor himself, on his identities and differences. These two notions have occupied a central place in the majority of counter-cultural movements—in the feminist movement and in those instigated by ethnic or national groups.

This calls for at least two criticisms. Firstly, the emphasis on identities and differences is not a new phenomenon. It is, as I have already pointed out, a characteristic of rising social categories and especially of new controlling classes, to assert their own identities and differences, in other words to demand freedom for their initiatives and the dismantling of traditional barriers which resist the advent of any new power. Secondly, there is an obvious contradiction between the libertarian search for identity and the desire to free this identity from the web of social relationships linked with work, because for a society to destroy its historicity and reduce its activity to the process of self-reproduction, it needs to exert strong social controls. Athens would have to become more Spartan than the real Sparta, preventing investment and the progress of knowledge, and reducing work to a series of acts which could easily become obligatory and ritualistic ceremonies. These criticisms are so obvious that the idea of outgrowing growth can only be linked to a very short period of our history, namely that when it seemed possible to abandon industrial society without abandoning abundance. This extremely brief period began in the early sixties, is already over in countries such as the United States, and is probably coming to an end in Western Europe. The study of student behaviour clarifies this point. Since about 1975, a plethora of counter-cultural models and social and political utopias has been brutally displaced by defensive behaviour and anxiety concerning professional prospects. This certainly does not mean that we should abandon criticism of industrial society, but rather that we cannot maintain very much longer the separation of cultural criticism from social and political criticism. Cultural criticism must become a criticism of the social forces which are leading towards a new type of society. Nor does this lead to a rejection of growth, but rather to the conception of ways for the collective reappropriation of the instruments and the products of this new growth.

A completely opposite criticism is implied in the notion that the idea of a programmed society is inadequate, since it still believes in the existence of a civil society and therefore of relations which can properly be called social. Critics of this persuasion maintain that such an image has corresponded to reality only in a very restricted area of the world and only during a very short time. This episode in civil society is in fact drawing to a close, and once again the state and the systems of maintenance and enforcement of social order are everywhere reasserting their domination. It is easy to perceive the reason for this criticism: the great hopes of industrial society were carried by the working class movement, and this movement gave the name "socialism" to the societal model opposed to capitalist society. Historically, however, this societal model which bore the workers' hope of freedom is embodied in a totalitarian state based on the reasoning of absolute power, and not the protection of the dominated classes. This is why the criticism of the communist model (which the more cautious would prefer to call Stalinist so as to limit its field of application) has led to the majority of industrialised societies being seen in terms of domination, in one way or another, by a state-imposed order. Instead of analysing new conflicts and new social movements, emphasis has been given only to repression, inculcation and restriction. This has in fact brought to light a certain number of social phenomena which had previously been hidden in the shadows; and which could severely distort social analysis to the extent that this picture of contemporary society appears to exclude the possibility of conflict.

However, reality is in complete conflict with this picture, leading to the conclusion that there is increasing disconnection between social relations, civil society and the state. It is true, and this is probably the most important phenomenon on a worldwide scale, that an increasing number of countries are engaging spontaneously in an ever more rapid process of industrialisation. The more spontaneous this process is, the less it is guided by social forces which have as yet been unable to form; the more, therefore, is it State-directed whether nationally or from another (foreign) State. It is in fact beyond doubt that today's world is far more State dominated than Western society was during the last century when undergoing the process of industrialisation. Yet once again, this phenomenon is not of the same nature as the passage from industrial to programmed society. It is wrong to speak here of a "power-elite", of monopoly State capitalism, or simply of a "total State". On the contrary it is more and more important to separate the analysis of the State from that of civil society. Even in a country such as France, have we not witnessed the development of the great, technocratic machineries taking place independently from political upheavals, from the weak Fourth Republic to the present industrialist State and including the truly Stately State of the Gaullist era, so much concerned with sovereignty or grandeur. The more weakly integrated civilian society is, the more it is made up of multiple networks of decision-making centres and fields of social influence, and the more the activities of the State are separated from those of society as such. For the State is concerned with historical change, and with maintaining the identity of a social entity caught between its past and its future and threatened by other surrounding entities; whereas civil society is an assembly of complex social relations, subjected at increasingly numerous points to conflicts and negotiations. The illusion of a possible return to equilibrium, like the illusion of an "all-devouring" State, can today serve only to delay analysis of the new directive forces, the new protest movements and of the interplay of their conflicts.

Conclusion

Thus there is not such a complete rapture between industrial society and whatever will follow it, as some have supposed. We shall not see the rebirth of "primitive" societies; nor shall we see social problems being absorbed within purely political issues, except perhaps in cases of extremely authoritarian states. The new society will be defined, just as were earlier societies, by its own process of self-modification which, through the medium of class relations, will evolve its own *modus operandi.* Production societies, which followed exchange societies, are in their turn being superseded by communications societies. What is radically new in these latest societies is that their capacity for self-modification extends to all areas of economic activity; they no longer see themselves as being subordinated to an order transcending social phenomena. All the preceding historical societies saw themselves in terms of an intermediate level somewhere between the realm of the Gods and that of Nature or of Demons. Merchant societies developed as a means of escaping extremism and of respecting the essential principles of civil life and peace. Industrial societies attempted to transform raw materials into historical progress. Programmed society can no longer perceive any superior order beyond itself, neither can it recognise the existence of a nature separate from itself. This is why it recognises on the one hand that it is both a part of nature and also responsible for nature's protection—that is to say it must undertake the management of all the likely consequences of the modifications it carries out on the natural order; and on the other hand it no longer recognises any god but itself, since it has the capacity to transform itself almost completely, or even to destroy itself. This is why society's understanding, for the first time, can no longer appeal to principles external to the social phenomena themselves. This had already been sensed by the first sociologists towards the end of the nineteenth century, notably by Durkheim, and is today becoming more and more obvious. We can and must study a communications society solely in terms of social relations. The meaning of the social actors' behaviour is not to be sought in principles, in an order of the universe, or in the current of history; it is purely and simply to be found in the social relations in which the actors belong. This is why society as a whole should be seen as a system of activities or, in the widest sense of the term, as a game. This also means that the analysis of society is, for the first time, necessarily sociological. Yet another result is that the mechanisms of social control or socialisation, since they can no longer be presented as respect for natural laws, or as the maintenance of traditional precepts, appear increasingly repressive. Everything that appears to be objective, established or institutionalised is found more and more, in this type of society, to be an impediment to the foundation of social relationships, an impediment to communication. This justifies the importance given to problems of the state in contemporary social thought. Programmed society is necessarily also a society of protest, imagination and utopianism because it is completely divided, on the one hand, by the social conflict of machinery which has the capacity and the power "to programme" and on the other hand by the call for creativity and happiness, which are constantly being threatened by the logic of the machines. This leads us to consider that such a society embodies a process of radical disinstitutionalisation, the most obvious victim of which is the educational system. Insofar as this system had a socialising function, through the transmission of a cultural heritage, it is today devoid of all meaning. As can be seen, the entire system is exploding: on the one hand, the new controlling classes are demanding that schools should become centres of training and preparation for professional life; on the other hand, there are those who oppose this process of professionalisation, on the grounds that it is part of a plan produced by those in power, but who would still like to see the demise of traditional school institutions, and make them into centres for play, creativity and communication. These two examples, the one concerning the understanding of society, the other concerning schools, show that the main area of confrontation is constantly getting closer to the realm of knowledge and ideas. This is explained not only by the fact that knowledge has become a productive force, whose appropriation by society is a problem as basic as that of industrial ownership, but above all by the fact that, the more a society extends its capacity for self-modification, the more it is guided by symbolic representations of itself and of social action, so the intellectual world, which had seemed to exist somewhere between society and the realm of principles and values, today finds itself at the very heart of social conflicts, completely unable to take refuge in the objectivity or the ivory towers of the scholars of earlier times. As a result this analysis of programmed society, rather than indicating a situation in which both knowledge and action should act simultaneously, points to a new relationship between knowledge and society. Sociology can no longer consist simply in bringing to light a social nature, it must analyse the component parts of all social objects so as to reveal the social relationships they hide, and thus the meanings concealed in what seem to be purely administrative or technical categories. It is in this sense above all that we must avoid all attempts to define this society as technological or to name it after certain of its technical instruments; this would be giving way to a partisan illusion—it would mean using a language whose partisan character lies precisely in its objectivity.

It is rather by seeing society as a field of conflicting social relationships (which is not to say that these conflicts cannot be resolved and cannot lead to relatively stable compromises) that we can most successfully approach the new characteristics of a society which no longer possesses a nature of its own, being entirely the product of self-generation, the result of its own action.

CHAPTER 8

Breaking the Chains of Traditional Systems Science

H. A. LINSTONE

THE systems approach is historically the brainchild of scientists and engineers. As such it is hardly surprising that it has inherited the paradigms and inquiring systems which contributed so much to the triumphs of modern science and technology. The stress is on compartmentalization and analysis, on data (or observations) and models. Even recent endeavors such as system dynamics*, cross-impact analysis, and technology assessment†, which deal with complex and multidisciplinary concepts, reflect this orientation.

Nine years ago the Dean of the School of Electrical Engineering at the Massachusetts Institute of Technology said:

> I doubt if there is any such thing as an urban crisis, but if there were, MIT could lick it in the same way we handled the Second World War.‡

He meant that the analytic approach which helped to develop radar and fire control systems should work just as well in solving

*__System dynamics,__ a science for understanding the dynamic action of social and other complex systems, developed by J. W. Forrester and others at the Massachusetts Institute of Technology.

†__Technology assessment,__ weighing scientifically beforehand the effects and social impacts, both favourable and unfavourable, expected to be brought about by actual applications of a new technology to be introduced or developed.

‡Quoted in W. I. THOMPSON, *At the Edge of History*, p. 48, Harper & Row, New York, 1971.

social systems problems. This attitude may be labeled self-confident by his friends and naive by his opponents. The scientist's belief in analysis is as strong as Billy Graham's in religion or Mao-Tse-Tung's was in Communism. The technology student is imbued with certain paradigms no less than the seminary student or party member; nearly all become convinced of the "rightness" and "truth" revealed in their texts.

Four Scientific "Truths"

Objectivity on the part of the analyst is vital for good systems work

The concept of the "objective" observer or analyst should be laid to rest once and for all*. If, in complex systems, "everything interacts with everything", then the analyst is related to his model or to the subjects of his experiment. This is particularly significant where human beings are elements in the system or its relevant environment—which is the case in virtually all interesting problems today. Mitroff and Blankenship have proposed guidelines for holistic experimentation, such as the following:

> The subject (general populace) of any potential holistic experiment must be included within the class of experimenters; the professional experimenters must become part of the system on which they are experimenting—in effect the experimenters must become the subject of their own experiments.
>
> The reactions of the subjects to the experiment (and vice versa) are part of the experiment and as such must be swept into its design (e.g., conceptualization)†.

From Adam Smith to West Churchman there have been expressions of concern with the danger of ignoring the individual, losing him in the system view. Smith said 200 years ago:

*cf. I. I. Mitroff, *The Subjective Side of Science*, Elsevier, New York, 1974.

†I. I. Mitroff and L. V. Blankenship: On the Methodology of the Holistic Experiment; An Approach to the Conceptualization of Large-Scale Social Experiments. *Technological Forecasting & Social Change*, vol. 4, p. 345, Elsevier, New York, 1973.

> The man of system . . . seems to imagine that he can arrange the different members of a great society with as much ease as the hand arranges the difficult pieces upon a chessboard. He does not consider that the pieces upon the chessboard have no other principle of motion beside that which the hand impresses upon them; but that, in the great chessboard of human society, every single piece has a principle of motion of its own altogether different from that which the legislature might choose to impress upon it.

Churchman paraphrases Kant:

> Two things fill my heart with never ending awe: the complexity of the total system and the self within*.

If the analyst himself has not focused on the process of individuation, i.e., understanding of the self, he will not be able to model social problems meaningfully.

Data and models form the basis for systems study

Man has always striven to understand the world around him and created models which reflect his perceptions to aid in that process. Modern man insisted that data are "objective" and that a model validated by data forms an excellent means to predict system behavior.

Physicists have already learned that apparently contradictory models may be useful in explaining a simple phenomenon (i.e., particle and wave theories for propagation of light). Furthermore, the choice of variables and data is, in fact, anything but objective. The use of GNP/capita as a measure of national status, the choice of variables in world dynamics models, and the definition of QOL (quality of life) are highly subjective. Measurability, i.e., quantifiability, is an important consideration in their choice†.

*C. W. Churchman: A Philosophy for Complexity. In: *Future Research: New Directions* (H. A. Linstone and W. H. C. Simmonds, Eds) p. 88, Addison-Wesley, Reading, Mass., 1977.

†cf. Gresham's Law of Systems Analysis; Quantitative analyses tend to drive out qualitative analyses.

The extensive use of computers underlines the tendency to stress quantifiable information. Workable models require structures, a very limited number of variables, and simple interactions among them. The modeler therefore abstracts the perceived situation until it fits this schema. The number of variables is reduced, nonquantifiable variables are scorned, and stochastic formulations are introduced. A simplified structure, preferably hierarchical, is postulated and nonlinear interactions are linearized. Heinz von Foerster's First Law expresses this process in another way:

> The more complex the problem which is being ignored, the greater are the chances for fame and success.

The usefulness of models is indisputable; it is their misuse which is deplorable. Like Pygmalion, the modeler falls in love with his creation and construes the artifice as a reality. At one time Euclidean geometry was accepted as the "truth", later it was Newtonian mechanics. One would think that we have learned from these experiences; yet, we make the same mistake with economics and world dynamics today.

The emphasis on data leads us directly to one of the problems in technological forecasting: the excessive reliance on trend extrapolation with its inevitable bias on the past, i.e., looking at the future in the manner of driving a car forward with one's eyes on the rear view mirror.

There is a best solution for a system problem

The formulation of a "problem" virtually presupposes the existence of a solution; a problem for which it is not possible to envisage a solution is usually accepted as a "condition". Most human beings fear death, but mortality is rarely identified as a problem; instead, it is accepted as a condition. Schools inculcate the idea that problems must have solutions: unsolvable problems rarely appear in textbooks.

Furthermore, there is a striving for the "best" solutions.

Classical mathematics have devoted vast resources to this goal (e.g., calculus of variations, game theory, and linear programming).

It is widely assumed that suboptimization is a valid process in complex problem solving. If a "best" solution cannot be found for the whole system, the suitable response is to find best solutions for the components or subsystems. Unfortunately, a set of optimal subsystem solutions may well constitute a very poor total system solution.

It is a fascinating exercise to construct a problem-solution relevance tree. Consider unemployment/underemployment as the problem. Viewing it in the context of the Maslow hierarchy we identify various aspects of this problem:

survival	—	minimum income
security	—	assured income and stable job
self-actualization	—	satisfying work

Taking any one of these we construct a relevance tree. The first level denotes the problem, the second possible solution concepts, the third indicates new problems raised by these solution concepts, and so forth.

It becomes evident that solutions alternate with problems *ad infinitum.* We immediately recognize that this exercise echoes the saying: "new technology creates new problems".

Problem: "assured income and stable job"

↓

Solution: seniority and tenure

↓

Problem: obsolescence of individuals and organizations

↓

Solution: education for change without generating insecurity (refocusing)

↓

Problem: psychic limits to change

This sequence is analogous in mathematics to an infinite divergent series and is similarly intractable, a situation which presents us with the one justification for applying a discount rate to future problems*. The discounting process transforms a divergent into a convergent series and permits us to compute the relevance tree, i.e., to determine preferences among solution concepts at any one level. However, Churchman argues that application of any positive discount rate is immoral†. If our children are more important than we are, we should be using a negative discount rate: a problem in the far future is thereby more significant than the same problem in the immediate future. Under this definition the orbiting of nuclear waste and any depletion of resources are immoral acts. Their effects on future generations should be viewed as more serious than the near term impacts since they appear to reduce the options of our grandchildren. So we are caught in a quandary.

If we *do not* use a positive discount rate, we become immobilized, unable to make decisions, like Hamlet we are paralyzed into inaction. If we *do* use a positive discount we appear to be immoral in the eyes of future generations. Our refuge is unbounded faith in our ability to brilliantly improvise a series of short-range decisions and to invent new options.

Specialization is essential in effective system study

The explosive growth of science and technology has been accompanied by a flowering of bureaucratic institutions. The organization and management of science have been accomplished by bureaucratic compartmentalization and specialization. A new subject of inquiry may originate as a naked, unlabeled topic, but it quickly assumes the garments of a discipline. The disciplining process occurred with engineering in the nineteenth century and

*For a discussion of discounting, see H. A. LINSTONE: On Discounting the Future. *Technological Forecasting and Social Change*, vol. 4, p. 335, 1973, and H. A. LINSTONE: Mankind at the Turning Point. *Technological Forecasting and Social Change*, vol. 7, p. 331, Elsevier, New York, 1975.

†C. W. CHURCHMAN, *op. cit.*

is happening to technology assessment and general systems theory today. The process is facilitated by the use of unique languages and jargons. Communication is easy within the discipline and discouraged with outsiders. Peer review of publications and financial resource allocation help to build a veritable Tower of Babel, protecting disciplines from each other.

Weinberg was acutely aware of this when he recently wrote:

> The general systems movement did not start out as a discipline but is probably ossifying into one*.

Unfortunately, the critical problems we must address do not fit the compartmentalization. They are not merely interdisciplinary, but increasingly transdisciplinary.

"We Shall Overcome"

These deeply ingrained "truths" illustrate the biases and illusions which enchain the systems approach and restrict its usefulness. The future does not lie with "more of the same". The preceding discussion alludes to some means to effect change and we will briefly outline them.

Develop dual foci for all complex societal systems: the total entirity and the unique individual

As Churchman has stressed, this does not mean that the unique individual is placed into the societal model†. Rather, it means that we must maintain two *Weltanschauungen* simultaneously, an inner and an outer view of the world (which may actually be in conflict). An understanding of the inner self affects the modeling of the outer world and this dialectic interaction is of distinct benefit to the task of dealing with complex systems. Individuation generates a spiritual, animistic awareness which forms the core

*G. Weinberg, *An Introduction to General Systems Thinking*, Wiley-Interscience, 1975, p. 46.

†C. W. Churchman, *op. cit.*

of new religious and cultural systems. It provides a nonrational, nonanalytic counterweight to the traditional "scientific" approach which is absolutely essential.

Use other inquiring systems

The value of a dialectic inquiring system was implied in the preceding paragraph. Churchman has stressed the potential of the Singerian inquiring system for interdisciplinary systems application*. It encompasses data based, model based, dialectic, and multimodel inquiring systems and provides the broadest possible modeling of any system†. It should minimize the tendency to get "hung up" on one model or one structure as happens so often. Singer views Truth as pragmatic, i.e., dependent on the objectives of the inquiry, and a system model is explicitly goal-oriented. Another useful concept is based on the work of Merleau-Ponty. A particular reality is created by "bracketing" an event or idea out of the mass of experience, truth is defined as agreement that enables action, and reality is the product obtained from intentions and actions‡.

Recognize that complex system problems have no "best solutions" in the classical sense

The search for "solutions" is not merely elusive, it is also illusive. We do not really solve complex systems problems, rather we *shift* or *transform* them. Forrester noted in his *World Dynamics* that action to reduce population growth would avoid a population crisis but stimulate a pollution crisis§. Avoidance of Mitroff's

*I. I. MITROFF and M. TUROFF: Technological Forecasting and Assessment: Science and/or Mythology? *Technological Forecasting and Social Change*, vol. 5, p. 126, Elsevier, New York, 1973.

†C. W. CHURCHMAN, *The Design of Inquiring Systems*, Basic Books, New York, 1971.

‡D. S. SCHEELE: Reality Construction as a Product of Delphi Interactions. In: *The Delphi Method* (H. A. LINSTONE and M. TUROFF, Eds) p. 43, Addison-Wesley, Reading, Mass., 1975.

§J. FORRESTER, *World Dynamics*, Wright-Allen Press, Cambridge, Mass., 1971.

Error of the Third Kind, i.e., solving the wrong problem,* must therefore be related to the concept of *homo ludens*—man as a game player.† We shift problems in accordance with our individual and societal needs or predilections (e.g., ethical standards, creativity, survival). Thus war has historically been a tragic, but acceptable, game in transforming problems. Science/technology represents another proper game—we will come back to its problem transformation capabilities in the next section.

Our love affair with the computer may well be traceable to its nature as a near-ideal game tool. A mature attitude toward holistic system design recognizes that (a) we cannot grasp any ultimate solutions to complex systems problems and (b) we need illusions as well as faith to survive. So it is with considerable humility and awe that we perceive the continuing evolution of human societal systems in spite of our ignorance.

Work for a better balance between analysis and synthesis or between reductionism and holism

I should like to conclude this discussion with an urgent plea. We can and must implement the principle that a system cannot be separated into components or subsystems without losing essential characteristics of the whole system. To do so we should now focus much more intensively on holistic approaches to system understanding.

Let us look at one facet—communication. The written language has provided the primary mode for technical communication since Gutenberg's time. Communication of nonlinear information in a linear mode obviously presents a handicap to most of us. There are those exceptional talents who have used the written word in a truly holistic way—we recognize the poet and novelist by their use of language to communicate emotions and a holistic "feeling" for a situation. Kafka and Orwell are

*I. I. Mitroff and M. Turoff: On Measuring the Conceptual Error in Large Scale Social Experiments: The Future as Decision. *Technological Forecasting and Social Change*, vol. 6, p. 392, Elsevier, New York, 1974.

†J. Huizinga, *Homo Ludens*, Beacon Press, Boston, Mass., 1955.

representative of this ability; science fiction writers generally are not.

It has long been understood that "a picture is worth a thousand words". Images have also successfully added the concept of time (i.e. television, motion pictures). It is now possible to study the design of a highway visually by use of computer-drawn time-variant images on a remote terminal. The designer can "see" the proposed road as if he were driving on it. Pilot training simulators provide a considerable degree of holism to aircraft landing and take-off operations.

The key is that the person is not a passive observer but an active participant in a dialog. Sensory and motor modalities are integrated to expand the information transfer rate enormously. The mutual reinforcement of multiple modalities is far more effective than an increase in precision of any one modality. And this proves to be feasible with computers used in a time interactive mode*. Another technique to enhance holistic communication is the concept of mind games. Developed by Robert Masters and Jean Houston, mind games facilitate altered states of consciousness which enhance images, subjective realities, and new space-time orientations†.

Jantsch suggests that holism can be approached in modeling by concentrating on the levels beyond cause and effect models (e.g., Newtonian mechanics) and statistically preprogrammed system models (e.g., econometrics). He reveals three higher levels:

(1) adaptive, evolutionary, morphogenetic dynamic models (e.g., role playing, ecological models, Prigogine's "order through fluctuation" principle)
(2) moral action models

*A. Johnson: Organization, Perception, and Control in Living Systems. *Industrial Management Review*, Winter 1969.

A. Johnson: Information Tools that Decision-Makers Can Really Talk With. *Innovation*, No. 10, 1970.

†R. Masters and J. Houston, *Mind Games*, Viking Press, New York, 1972. A workshop in our Systems Science Doctoral Program at Portland State University provides training in this approach.

(e.g., learning games, inventive system models, "live" simulation)

(3) creativity or *Zeitgeist* models
(e.g., new cultural paradigms, generation of myths, religions, ideologies).

We cannot expect to be successful at all these model levels, but they point the way toward holism in human system design. A small, but instructive, clue is provided by a recent study of the Parthenon, long considered a model of symmetric perfection. It was found that this edifice was built for the human eye, not the calculating brain. Departures from uniformity were intentional, yet "deliberately casual and designedly unsystematic, having been taken at random for a purely aesthetic purpose, in order to temper lifeless mathematical rigidity with those minute irregularities which distinguish the living organism from its abstract generic pattern"*. This adaptation of technology to man clearly represents modeling at a more holistic level than practised by traditional scientists and systems planners.

Medicine provides another illustration of the insights gained by holistic thinking. In the United States research has focused with considerable success on the conquest of disease. Organ replacement and life extension are popular fields of activity. Today there are twenty-two million people aged 65 and older; in 2000 there may be twenty-eight million. As medical advances increase life expectancy, the retirement age is moving down. A man retiring in 1970 at age 65 could expect to live another 13 years; a man retiring in 2000 at age 55 can expect to live another 28 years. Thus the retirement period is increasing more than 100% in one generation! With childhood and school attendance growing longer we find that a man's life will consist of three roughly equal periods: 20 to 25 years of youth activities, 30 to 35 years of "work", and 25 to 30 years of "retirement". While the biological/physiological aspects of the aging problem have been addressed in strength, sparse effort has been devoted to the social/

*R. Carpenter, *The Architects of the Parthenon*, p. 14, Pelican Books, 1970.

cultural/economic phases. The quality of life and function in society for the aged deserve equal attention in holistic systems design. We are beginning to search for a balance among the dimensions of life maintenance, quality of life, and societal role for the last third of the life span.

Communications and transportation technologies have shrunk the earth drastically in the past 50 years. We find, however, that the movement of goods and people requires far more energy than the movement of ideas. There are, therefore, trade-offs between transportation and communications which holistic systems analysis is starting to take into consideration. The impact of these technologies on the human psyche itself is of profound significance. The average American household spends 8 hours per day watching television and the average individual spends 27 hours per week. Children raised in a television era become accustomed to passive entertainment; Jean Houston refers to the "switching society"*. With one channel depicting news footage from a Middle East war and another a science fiction story, the distinction between fantasy and reality is blurred for the young child. Sensory overload and inundation lead to an inability to cope and an urge to drop out. The high degree of mobility not only means a break-up of the multigeneration family but higher degree of discounting of the future (and past) by the individual. The rootless offspring is not interested in maintaining the continuity of home and tradition or of family history. We also observe an increasingly overstressed and overtranquillized society. Tranquillizer sales have been growing at an annual rate of 9%.

Finally, worldwide cultural homogenization and creeping cyberveillance† of individuals are possible consequences of communication capability which give rise to grave concerns.

Like a child playing with fire we find that the game of communications/transportation technology may have extremely serious unintended impacts. Our concepts of morality may well

*J. HOUSTON: Prometheus Rebound: An Inquiry into Technological Growth and Psychological Change. *Technological Forecasting and Social Change*, vol. 9, No. 3. Elsevier, New York, 1976.

†Cybernetic surveillance.

dictate a change in the rules of the game.

We have here clues that a truly holistic system approach should be fruitful for environmental synthesis. Obviously, we are still very far from achieving a *balance* between reductionism and holism, analysis and synthesis, yin and yang. But that is the hope which nourishes us.

Postscript

The preceding article was prepared for the Tokyo Discoveries Symposium in 1976. The author is now (1979) developing the principal ideas discussed above in the context of a project sponsored by the National Science Foundation on "Alternative Models of the Decision Making Process and their Applications to Technology Assessment"*.

In 1971 Allison† demonstrated the advantage of using three perspectives in an analysis of the Cuban missile crisis: the rational actor or quasi-technological approach (Model I), the organizational process (Model II), and the bureaucratic politics or individual focus (Model III). These involve entirely different paradigms, not merely different mathematical models. The current project contends that a crucial weakness of technology assessments has been the emphasis on Model I and neglect of the other perspectives. Mini-case studies suggest that the Allison models could also provide remarkable insights for technology assessment which cannot be elicited from Model I. We hypothesize that (1) the Allison models can significantly upgrade the usefulness of technology assessments for decision making and (2) the balance among the models varies with technology type.

Models II and III are being applied to several technology assessments and the results will be compared to those obtained with the more traditional systems-oriented Model I. The project will be completed in February 1981.

*H. A. Linstone and W. H. C. Simmonds (Eds), *Futures Research: New Directions*, pp. 136–39, Addison-Wesley, Reading, Mass., 1977.

†G. T. Allison, *Essence of Decision*, Little, Brown, Boston, Mass., 1971.

CHAPTER 9

Individuals and Society

T. TERANO

Introduction

Human society has developed in order to protect the interests of its members, but it often happens that individuals' interests are adversely affected by societal constraints. Attempts to resolve this conflict have been made since ancient times in the fields of the humanities and social sciences, but no real solution has yet been attained. Indeed, modern theories of social science deny the existence of a democratic way of "social choice"*.

The rapid developments of technology have nevertheless enhanced the awareness of rights on the part of individuals. As a result, the conflict between individuals and societies, or between groups within a society, are becoming more intense than ever before. If this trend should continue, the world will be ruined by disputes over societal philosophy long before material shortages become critical.

At this juncture we scientists, rather than "pursuing the truth" regarding all problems existing between individuals and societies, ought to be working out rational rules for reaching compromise in the more outstanding of these problems. Thus we should be able, reasonably rapidly, to alleviate the concrete practical difficulties —using the social sciences, in a goal-oriented fashion, for synthesis rather than analysis. In this sense, our social sciences should more correctly be classified as engineering rather than "pure science". According to my own definition the science of social systems, in fact, is nothing other than the application of the

*Refs (1) & (2) p. 167.

methodology of systems engineering to the various problems of social science.

Engineering methodology is superior in terms of its goal-orientation, rationality, precision and positivism. If this type of methodology could be applied to the problems of social science, it would without doubt become a very powerful tool in that field. However, the objects of engineering and those of social science differ widely. For example, in the recognition of an object in engineering, there are practically no individual differentials and it is not so hard to work out generalised theories. On the other hand, in those fields dealing with human societies, the objects are too complex to establish even a "basic cognition of objects", and there must therefore be a variety of interpretations. It is also very difficult to prove anything by measurement or experimentation. Therefore, if engineering methodology should be introduced without fully considering assumptions of data which are the premises of theorisation, there is a danger of good intentions leading to wrong conclusions.

After all, in the methodology of dealing with human beings or societies, the analytical processes must be generalised and objective. But the conclusions derived from them must be elastic, with concordant width and depth. In particular, in order to deal with the claims of individuals and the demands of society simultaneously, it is necessary to develop new methodologies which can embrace the inherent multifacetness, self-contradictions and ambiguities involved. As examples of such methodologies, I would like to describe "Fuzzy Set Theory" and "Structural Modelling". The former is appropriate for analysis of individuals' behaviours and thinking; the latter for analysis of societal systems.

Analysis of Subjective Evaluations through Fuzzy Set Theory

Set theory is a field of mathematics which presupposes a group of elements with similar qualities and discusses the relationships among such groups. A Set is a concept which is very abstract, qualitative and macroscopic. L. A. Zadeh* has expanded the scope of Set theory and developed theories to deal with an ambiguous boundary (Fuzzy Sets). These are very flexible and comprehensive, in contradistinction to more conventional mathematical expressions which are assertive and "unifaceted". Set theory does not presuppose infinite trials, nor focus upon average values, as does probability theory. Therefore it is convenient for expressing subjective ideas and behaviours of human beings. For example, the concept of *Young Lady* differs widely from one individual to another, and is thus very ambiguous. One way in which it may be expressed is shown in Fig. 9.1, where age is indicated on the horizontal axis, and "Grade" of belonging to the *Young Lady* category (Set) is indicated on the vertical axis.

Fuzzy Set theory has a relatively short history. Therefore its applications are not yet very well developed. However, it is attracting attention in such fields as artificial intelligence,

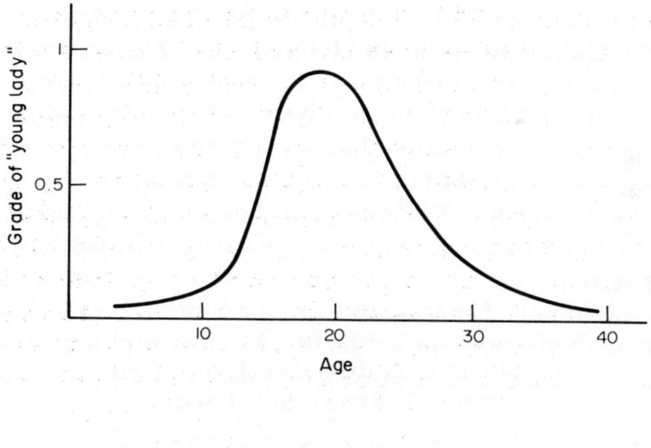

FIG. 9.1 An illustrative example of a fuzzy set

*Ref (3) p. 167.

behavioural science, sociology, etc*. It is expected to bridge the gap between engineering on the one hand, and the humanities and social sciences on the other. Here I would like to indicate, as an illustrative example, a mathematical model for subjective evaluations by individuals.

In contemporary societies many problems have their roots in the gap between individuals' evaluations and the average evaluations of a society. The difficulties involved in evaluation problems derive from two main causes. The first is the fact that the criteria and the numerical scales of evaluations are not clear: we rely exclusively on the sensitivities of the evaluators, which are difficult to quantify. The second cause of problems is that the objects of evaluation usually consist of many parts, and the values of evaluation for each component must be synthesised to get evaluation scores for the whole. The degree of importance which the evaluators have for each component differs from one individual to another. Also, the method of synthesis varies. Accordingly, even if the evaluation scores of each component should be identical, the synthesised evaluation scores could differ widely from one individual to another.

Let us consider, as an example, evaluation of a human face†. A human face consists of such parts as the eyes, nose and mouth; in other words, many components for evaluation. The evaluation scores of these parts are bound to be uncertain for the reasons just indicated. Therefore we may prefer to express them in terms of a "Grade"—a representative quality belonging to a Fuzzy Set of *Beauty*. On the other hand, in evaluating the face as a whole, the subjective preference which the evaluator has for each part can be expressed in terms of the "Grade" of ascription to a Fuzzy Set of *Relative Importance*. In this overall evaluation, we must make an overall judgment of ambiguous amounts of ambiguous criteria.

As a mathematical model of this kind of problem, the Fuzzy integral developed by M. Sugano‡ is most suitable. We divide the

*Refs (4, 5) p. 167.
†Ref 6 p. 167.
‡Ref 7 p. 167.

human face into five parts—the eyes, the nose, the mouth, the chin, and the profile. These five parts are illustrated in Fig. 9.2.: they may be denoted by the Set

$$Si(i = 1 \sim 5).$$

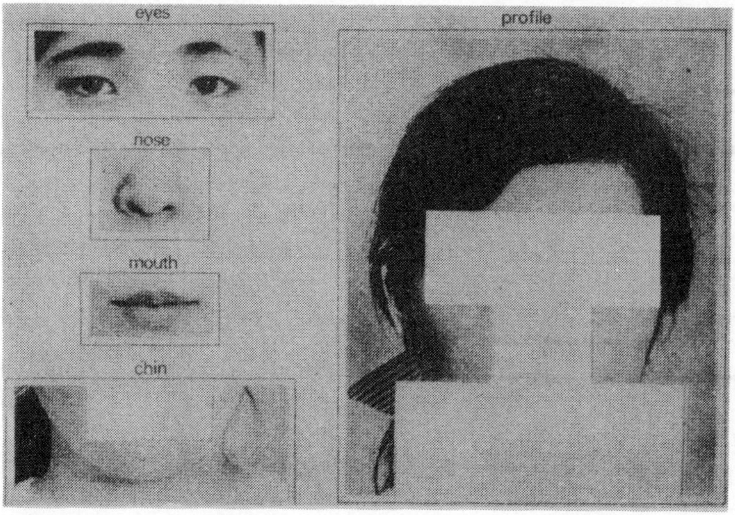

FIG. 9.2. Evaluation of facial components

If the evaluation score of each part is expressed by $h(_{si})$, and the relative importance by $g(.)$, the Fuzzy integral is defined by the expression

$$f\Lambda h(s) \cdot g = \overset{n}{\underset{i=l}{V}}[h(si)\Lambda g(A \cap Ki)]^* \qquad (1)$$

If we compare this with the expression $\sum_{i=l}^{n} h_i \times W_i$, which is a

*In this formula, f stands for fuzzy integral. V and Λ are operators which select max. and min. respectively. A stands for integral area and Si is numbered in the order of $h(s1) \leq h(s2) \leq \ldots \leq h(s5)$.
$Ki \triangleq (Si + 1, Si + 2, \ldots S5)$

weighted mean formula conventionally employed, we can see that $g(.)$ replaces the weight W_i and the operation minimum and maximum correspond to multiplication and addition respectively. In other words, the expression (1) represents a macroscopic calculation of expected values.

In this expression $g(.)$ satisfies the following equation:

$$g(si \cap si) = g(si) + g(si) + \lambda g(si)g(si) \quad -1 < \lambda < \infty \tag{2}$$

This indicates that the weight of overall evaluation is not merely the total sum of the partial weights. Rather, the cross effect between these components should be taken into account. If λ is positive, this means that the combination of the components is important—if negative, it shows that there are some duplications among the partial evaluations.

The author prepared a picture similar to that shown in Fig. 9.2, and in an experiment asked students to make partial and overall evaluations. We can determine the values of $g(si)$ and λ so that the result of calculation using the expression (1) agrees with the experimental scores as much as possible. Then we can tell a student which part of a woman's face is most important for him.

Thus we have achieved at least a partial mathematical model of the student's mind. Fig. 9.3 shows an example, and Fig. 9.4 shows the experimental values and the calculated values of synthesised evaluation in this case.

All this confirms that a Fuzzy Set is a very ambiguous mode of expression. However, it seems to express essential attributes better than precise mathematical formulae in providing a model for complex human characteristics.

Social Structural Model

Another method for expressing multifaceted systems qualitatively is a structural model. In this, we take into account very basic elements comprising the system, and their inter-relationships expressed by a network graph. If we select an appropriate

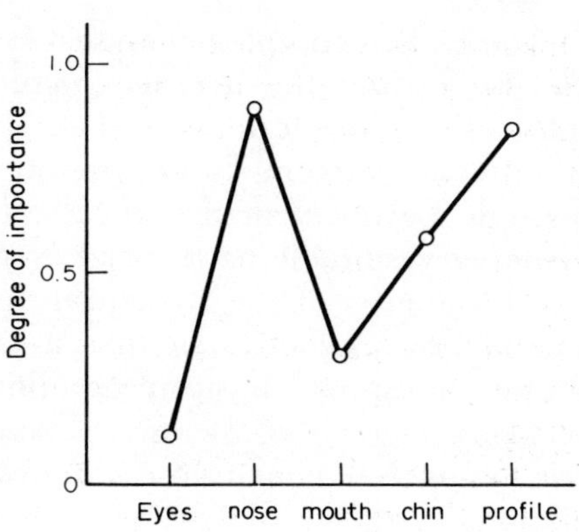

FIG. 9.3. Degree of importances attached to each part

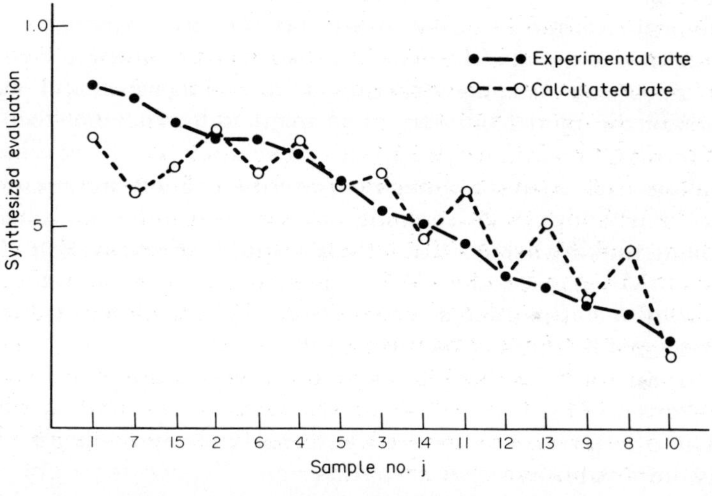

FIG. 9.4. Comparison of calculation and experimentation in the appraisal of a human face

number of elements, it is possible to express concisely and exclusively the essential structure of any complex system.

A structural model can be made exclusively by experiences and intuitions of the model builder without having to deal with the numerical data or to conduct complex analyses. It is also possible to make a model by coding the average opinions of a great number of people, using questionnaires. The structural graph itself is very qualitative. Therefore its interpretation has latitude. The more macroscopic a given structure, the greater the latitude. In this regard, it is very suitable as a social system model. To further enumerate the merits of the structural model, it can be intuitively understood by anybody, so that it is suitable as a means for obtaining widespread basic recognition. If the interrelationships between elements are given quantitatively, it immediately leads to concrete conclusions through simulation or numerical calculations.

As a method to establish a social structural model, DEMATEL* and ISM†, both developed by Battelle Research Institute, are most interesting. Neither of these employs an analytical technique: they make people use experience and intuitions exclusively. The characteristic feature of these methods is to make manifest the structure of consciousness latent in the minds of the model builders. The major difference between the two is that the former deals with the structures of world-wide problems. Thus, questionnaires are sent to the influential people of the major powers: by so doing, it is hoped to work out solutions through making visual the structure of consciousness on their part. ISM on the other hand concerns itself with immediate everyday social problems, and tries to explore optimal policy by making explicit the structure of the problems through group discussions and learning among those who are affected by the problems.

These methods are effective when problems are so complex that it is impossible otherwise to determine where to begin their analysis or to work out countermeasures, or when there are divergencies of views among the several parties involved so that the differences in their understanding or evaluation of the

*Ref 8, p. 167.
†Ref. 9, p. 167.

problem, which are the cause of these disagreements, are not clear.

As an illustration, let us consider the structural identification of the complex factors present in the local administration of a city.

First, we select three hundred keywords from matters concerning local administration. When these keywords are classified and codified, we obtain the following 22 items.

(1) Improvement of electricity, gas and water service
(2) Promotion of industries
(3) Vocational training
(4) Promotion of education and culture
(5) Provision of good leisure facilites
(6) Problems related to sex
(7) Demographic problems and problems resulting from age distribution
(8) Problems related to overdensity or sparsity within the population
(9) Pollutions
(10) Protection of the natural environment
(11) Prevention of disasters and crimes
(12) Peace of mind
(13) Health and welfare
(14) Integrated development of transportation
(15) Comprehensive plans to cope with social problems
(16) Making municipal finance sound
(17) The residents' participation in local autonomy
(18) Dissatisfaction with lacunae in legal systems
(19) Regional egoism (attempts to satisfy the self-interest on the part of residents in a given region in a self-serving manner)
(20) Urban renewal
(21) High quality housing construction
(22) Rationalisation and streamlining of distribution processes.

The second step is to consider these items as the elements of a structure, and to discover their inter-relationships. For this purpose any two given items are selected, and examined to determine whether or not causality exists between them. If there is a causal relationship, the items are connected with an arc (this judgment must be done intuitively). In this way we can obtain a directed graph where vertex represents element, and arc represents inter-relationship.

With the twenty-two items mentioned above, the structure to be represented is highly complex, and difficult to represent in the graphical form described. It is therefore represented in a *vertex matrix*, as in Table 9.1, rather than a graph.

	1	2	3	4	5	6	7	8	9	10	11	12	13	14	15	16	17	18	19	20	21	22
1		1							1	1		1						1	1	1	1	
2			1					1	1	1						1				1		
3							1															
4					1							1										
5						1				1		1	1									
6							1					1						1				
7	1		1	1	1			1					1	1	1					1		
8									1		1			1					1	1		1
9										1	1		1					1	1	1		
10					1				1		1	1	1									
11												1										
12				1																		
13											1	1								1		
14								1				1										1
15	1	1		1	1			1	1	1			1	1		1				1		
16		1												1						1		
17									1	1	1	1	1						1		1	1
18																	1		1			
19	1								1	1				1	1			1		1		
20	1			1	1				1	1	1			1		1					1	
21											1	1										
22								1			1			1								

TABLE 9.1: Vertex matrix dealing with urban problems

In this matrix for example, the fact that the element of column 9 of the line 1 is *one* indicates that unless good utilities are provided, "pollution prevention" is impossible. A blank space signifies that the element is zero, that is, the inter-relationship is weak.

Because such a structure, due to its great complexity, does not lend itself to graphic representation, it can be useful to further reduce the structure. To do this we may enlist the help of the "Bottleneck Method"*. As a result of such an operation, the items which are closely inter-related are combined to form subsystems, thus simplifying the whole structure to the form shown in Figure 9.5. If we consider the physical meanings of the subsystems, and integrate them into still larger subsystems, an even more simplified structure is finally obtained, as shown in Figure 9.6.

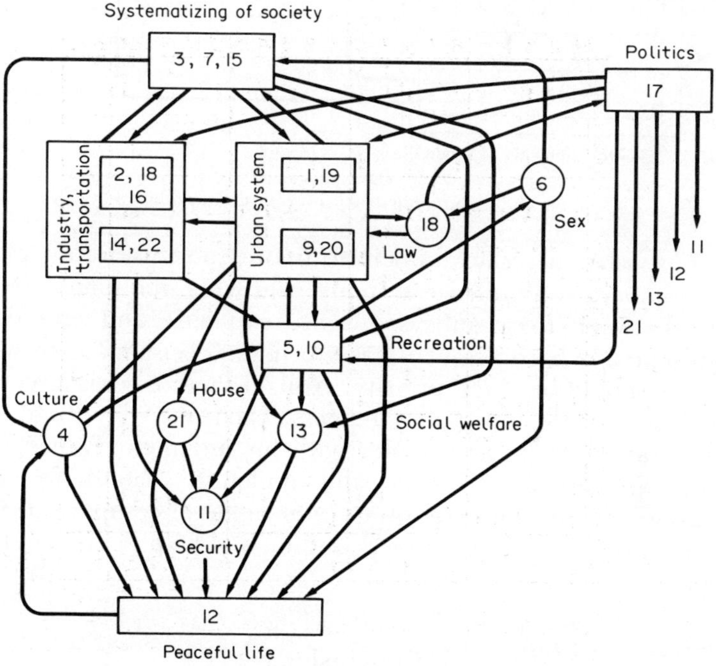

FIG. 9.5. Structure of urban problems

*Ref. 10, p. 167.

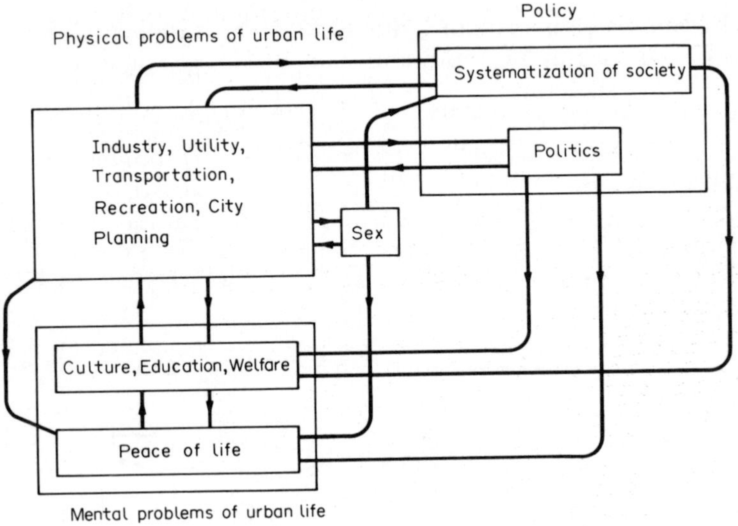

FIG. 9.6. Simplified structure of urban problems

Thus, when a great number of minor problems are integrated into comparatively few subsystems, and when the inter-relationships between these subsystems are explicitly and graphically represented it becomes very easy to make decisions as to which policies should be given priority. Also, if there is a disagreement between two or more policy makers, each one may build up a structural model in accordance with his judgment. When these models are compared, it is possible to identify visually the causes of disagreement. This use of a visual mode of comparison must certainly deepen mutual understanding and help to prevent futile disputes.

Conclusions

If we hope to solve the problems between individuals and societies through scientific methodology, it is imperative in the first place to create a common understanding. The two theoretical approaches here described can be of great future use, due to their flexibility, as methods for understanding human behaviour.

As indicated at the outset, since systems science is a goal-oriented discipline, it should not content itself with analysis of problems. For this purpose all the methodologies now being developed (e.g. simulation methods, technology assessment methods, evaluation techniques, systems analysis methods, decision-making methods) should be employed in order to provide some concrete solutions.

I believe this is possible only through interdisciplinary cooperation.

1 K. J. Arrow, *Social Choice and Individual Values*, John Wiley, New York, 1951.
2 A. K. Sen, The Impossibility of a Paratian Liberd, *J. Political Economy* **78**, 152–57, 1970.
3 L. A. Zadeh, *"Fuzzy Sets", Information and Control*, vol. 8, pp. 338–53, 1965.
4 L. A. Zadeh, (Ed.) *Fuzzy Sets and Their Applications to Cognitive and Design Process*, Academic Press, New York, 1975.
5 M. M. Gupta (Ed.), *Fuzzy Automata and Decision Processes*, Elsevier, New York, 1977.
6 H. Kato, *Studies of Evaluation of Women's Faces*, Master's thesis, Tokyo Institute of Technology, 1973.
7 M. Sugano, *Theory of Fuzzy Integrals and Its Applications*, Doctoral thesis, Tokyo Institute of Technology, 1974.
8 E. Fontela and A. Gabus, *DEMATEL (Structural Analysis of the World Programatique)*, Battelle Report, Geneva, 1974.
9 J. N. Warifield, *Toward Interpretation of Complex Structural Modelling*, IEEE, SMC A-5, 1974.
10 T. Terano and N. H. Quan, *Macroscopic Structural Analysis*, The 2nd Systems Engineering Symposium, SICE, 1976.

CHAPTER 10

Impacts and Characteristics of Automation and Automated Information Processing Technology

R. SHIRANE

Genealogy of Automation

Technology is a means to assist, substitute and expand various aspects of human activities. Aristotle stated that technology may be divided into two categories: that derived from the spirit of labor—or the needs of real life—and that derived from the spirit of play. According to this classification, automation is a typical technical system evolved from the spirit of labor.

Since any attempt to trace the origin of automation technology necessitates going back to the origin of interaction between man and technology, the author has limited himself to seeking the genealogy of automation in the development of modern technologies.

The first move toward automation was the introduction of automatic machines into production systems, which achieved rapid progress in the 1920s and 1930s. One of the driving forces in the mainstream of this trend was the development by Henry Ford of the mass production system for automobiles, which brought about a great innovation in industrial production. With the idealistic slogan: "The purpose of an enterprise should not be just to make a profit, but to serve society", the flower of Fordism bloomed. Amidst intensifying conflicts between labor and management, in a situation of severe working conditions and low wages, Ford proposed the double and at first sight paradoxical target of low prices and high wages.

The supply of cheap good quality automobiles while paying high wages to workers were clearly two objectives of great benefit to the community; but objectives which would have been mutually exclusive without the establishment of a mass production system by automating the manufacturing processes.

Automatic operation and control of casting machines and of machine-tools for various functions was therefore introduced. In response to the demand for interchangeability of parts, the "go/no-go gauge" and photoelectric cell were developed for hardware control, while in the software field standardisation and statistical quality control systems were initiated.

In spite of these advances, however, the systems introduced at this time for control and automatic operation remained very primitive and far from what is now called automation.

Many technical barriers to further progress were breached as a result of the rapid developments in World War II. For instance, the development of servomechanisms for the automatic control of anti-aircraft artillery, and of the electronics needed to upgrade the sophistication of these systems, made great strides forward as a result of radar development. While, in fact, the incentive for technical development had been said to depend upon either the spirit of labor or the spirit of play, it cannot be denied that the progress of automation was accelerated by the spirit of struggle, as was the case with other important technical systems. Thus, toward the end of World War II, it became possible to employ automatic controls and operations in far wider areas than previously, causing a spin-off of industrial production systems and launching the so-called automation revolution.

The innovation of this revolution in the production process was to substitute the function of operators with automatic control devices, which had not previously been possible with individual automatic machines. Now, however, the machines were made to monitor the results of their own work, to decide what to do next and to manage the overall processes. This suggests that the operator's work, in its previous sense, could no longer be necessary. A typical example is transfer-type automation uniting the transfer machine lines in an automobile plant, now regarded as representative of mass production assembly plant based on modern machine processes.

On the other hand, since the latter half of the 1950s, a new direction in process automation has been the development, in the United States and Britain, of *program-control* systems. While the *transfer-type* automation is applicable only to mass production, program control can be applied in small-scale production. That is, the control equipment is actuated by commands from punched tape or magnetic tape, and a general-purpose machine can thus perform a changing variety of work without human intervention, so realising a variable and flexible production system. Moreover, the innovation inherent in the concept of variable and flexible program control is very deep and extensive, promising the development from the automation revolution to the cybernation revolution.

Automation and Cybernation

"Cybernetics", published by N. Wiener in 1948*, established a very important basic concept for automation. As suggested by the subtitle of his book, "Control and Communication in the Animal and the Machine", Wiener maintained that, if communication and control functions were taken into consideration, the same analytical principles could be applied both to machines and to biological systems. This means that, by considering information flow and feedback, automatic control theory can be applied not only to conventional automatic machines and production systems, but also to social systems and bio-engineering. Some specialists have called this development cybernation (derived from automation and cybernetics) and regarded it as a new innovation well beyond conventional automation.

The initial purpose of program-control type automation was to

*N. Wiener, *Cybernetics—control and communication in the animal and the machine*, John Wiley, New York, 1948.

produce "one-off" parts or simple assemblies in the machine production system. However, as the digital computer was introduced to the information control function as an automatic information processing system, functions of more advanced and complicated information processing and management were added, to expand the range of applications enormously.

One direction in this development is to combine individual automated machines into an integrated and coordinated system, including functions of both production control and quality control, with the ultimate aim of developing a completely unmanned plant.

Another direction (which seems more realistic) is the application of process automation to various other fields, and this has already achieved some success. As objects of program control in a production system, machining and assembly lines do in fact present some rather difficult problems, and it was to be expected that the techniques would be more readily adopted where their objectives were more easily achieved, such as in processes for handling electricity, gas and fluids. For instance, highly advanced automation systems controlled by computers have already been realised in power generation/transmission/distribution networks, in petroleum refineries and in petrochemical plants.

Such applications have recently been further extended to solid-handling processes, for example the most modern steel production plants. This type of automation now extends beyond the boundaries of industrial production (secondary industries), on the one hand into the fields of primary industries such as agriculture and mining, and on the other hand into tertiary industries such as transportation, communications, and the banking system.

These developments are characterised by the gradual shift of applications from industrial fields to non-industrial social systems such as administration, education and medical care. Generally speaking, it is more difficult for the principles of efficiency to penetrate into these non-productive systems which, partly due to their long history, are comprised of subsystems which have become inflexible vertical divisions. Bureaucracy itself is a typical example of this, having lost its flexibility through restrictive regulations, traditions and customs, until it has become a system of mechanically rigid structure, not capable of adapting to environmental changes.

If these rigid subsystems are combined horizontally, and Wiener's functions of communication and control are incorporated, each subsystem can be reorganised in accordance with the overall purpose of the total system, and flexibility to adapt to environmental changes may be regained. In this way the social system may be revitalised—that is, the system may be revived by cybernation.

A good example is the achievement of the modern Japanese regional emergency medical information system.

Each regional social service subsystem for emergency medical care consists of hospitals and clinics in the region, fire stations which operate ambulances for emergency patients, blood banks, the appropriate authorities of the local municipality, and the regional residents, who are taxpayers and beneficiaries at the same time. The emergency medical information center in each region combines these components organically so as to provide fast and appropriate treatment to victims of sudden illness or traffic accidents.

This life-saving medical care service exemplifies the great advantages offered to residents, as service-beneficiaries, by the introduction of cybernation into the social system.

The Computer as an Automatic Information Processing System

As automation and cybernation become more sophisticated and complicated, the computer's function in information processing is assuming increasing importance as an addition to its communication and control functions. This is undoubtedly one of the most important developments of the century, the significance and predicted future development of which will now be discussed.

The computer and innovation

The computer is a technological system offering many

important possibilities for innovation in information processing. Most of these possibilities have yet to be exploited. For example, in addition to calculation, data storage, rearrangement and retrieval, it becomes possible to derive logical conclusions of events (provided they can be described by binary logic) and to present the results in any desired format.

The information activities of human beings are both complicated and multifarious. They may be classified in accordance with five functions: (1) acquisition, (2) storage (memory), (3) transmission, (4) processing and (5) creation.

Various systems for the technological simulation of each of these activities (except for creation) have already been realised, some of the more important being:

(1)	Acquisition	radar, electron microscope, remote sensing.
(2)	Storage (memory)	various printing techniques, magnetic recording, semiconductor integrated circuits.
(3)	Transmission	telegraph, telephone, radio, television.
(4)	Processing	computer systems.

It should be noted that the last of these, (4: information processing) had been considered beyond the coverage of any technological system until the development of the practical computer in 1946. That is to say that, while other fields have a long history of technological development, the introduction and application of the computer as an information processing device—with a history of only 30 years or so—seems to have only just begun. Recognising this is important.

Furthermore, the introduction of computers, still largely an adventure in unexplored territory, is continuing, and continues to have an intensive impact, having already begun to affect every aspect of our life either directly or indirectly.

The industrial revolution, symbolised by the invention of the steam engine, started in the mid-eighteenth century in England.

The technology of mechanised production was subsequently established through the development and utilisation of electrical energy.

This stimulated the "consumption revolution", changing the so-called agricultural society into its modern successor, the highly industrialised society of today. The advent of the computer 200 years after the industrial revolution is expected to provide the motive force for the "information revolution" that is, conversion from an industrialised society to an information society. This is a common way of describing the innovations expected from computer technology.

It should be noted, however, that for evaluating these innovative developments of computer technology, its excellent information-processing capabilities (including superhuman operational speed) tend to be overemphasised, ignoring another important function—namely the expansion of storage (memory) capabilities.

Above all, the increased capacity and reduced costs of memory elements using modern semiconductor integrated circuitry are contributing greatly to the expansion of the computer's capability and its fields of application.

In human informational activities, "a sure memory" is one of the most outstandingly weak points. A most important role of computer technology is to compensate for this and so expand human activities, meaning that the computer's capability to provide "sure" storage and retrieval of information is a great merit.

On the basis of these facts, the innovative promise of the computer may be expressed in the following way: if the human brain, which is strong in analog information-processing such as pattern recognition, is combined with the computer's high-speed and reliable performance of complicated digital logic processsing to form a mutually complementary man-machine system, the performance of man's information-processing activity can be markedly improved and the evolution of a "new human being" realised.

Development of computers

It is generally accepted that the first practical electronic computing machine was ENIAC (Electronic Numerical Integrator And Calculator) completed in 1946. This machine was developed by Professors J. P. Eckert and J. W. Mauchly of Pennsylvania University under the sponsorship of the U.S. Army for the purpose of ballistics calculations. However, ENIAC was based on a rather different system concept from that of modern computers, the stored program system proposed by J. von Neumann in 1947. The first machine using this later system was EDSAC (Electronic Delay Storage Automatic Calculator) completed in 1949 in England by Professor M. V. Wilkes at Cambridge University. EDSAC can truly be said to have paved the way for the development of today's computers.

The subsequent development of computers is described in many accounts elsewhere, and no further mention is made here. However, it may be noted that continuing technical advances have led to operating speed, memory capacity and reliability being improved ten-fold while the cost of a machine has been reduced to one-tenth every 6 years. Following computers of the first generation using electromagnetic relays and vacuum tubes, and those of the second generation using parametrons and transistors, the present computers of the third generation employ mainly semiconductor integrated circuits (IC); and with the great improvement in the degree of integration following the advent of LSI (large-scale integration), the so-called 3.5-th generation is beginning.

Concerning the fourth generation, because basic design concepts of the computer may be changed extensively in the near future, there is as yet no clear image. Future developments are likely to concern the following:

(1) *Increase of processing speed.* As the degree of integration increases with LSI used for logic elements, processing speeds several times faster than those of third generation systems are expected to be achieved, resulting in both

high reliability and low price. Further improvement in this direction would greatly extend the application area of computer systems.

(2) *Introduction of "firmware".*
With increasing size and performance of computers, the software also becomes increasingly complicated and vast. Recently, software has tended to be partly incorporated into hardware to make up "firmware". This allows the "modularisation" of software, leading to its production in the factory and thence to increased operability of the whole computer system.

(3) *Progress of language.* In order to make a computer readily usable for everyone, it is desirable that its operational language should be as similar as possible to normal languages. End User Language (EUL) may be cited as an example of the conversational-type languages developed for this purpose.

(4) *Development of terminal equipment.* A basic field for improvement concerns ease of handling and performance of input/output devices, which are located at man-machine interfaces. The development of pattern recognition technology and various types of "intelligent terminals" is being actively pursued.

(5) *Network construction.* Computer operating modes are progressing from simple on-line applications to networks of multiple computers. This allows the combination of hardware, software, and data bases into common resources, enabling independent computers to act as an integrated organism. In line with this tendency in computer network architecture which will permit the interconnection of computers of different natures, the preparation of corresponding data bases is in progress.

Complexing of computer technologies

Besides developments in the computer system itself, as just described, it is also useful to have a general overview of developments achieved by combining (complexing) computer utilisation with other related technologies.

A first area of interest is the rapid expansion of computer applications through recent advances in communications technology. These have led to the transition of computer system form, first from point to line, then further to extensive plane-like networks. Recently, commercialisation of digital data exchange networks suited to computer communications has begun in advanced countries, so accelerating computer utilisation. Moreover, the use of satellite communications is rapidly advancing not only internationally but also in the field of domestic communications, and with the practical application of optical communications systems, capable of transmitting great amounts of information, substantial reductions in communications costs are expected. This will further extend the application range of computers.

One of the directions for future development is the application to region-wide social services, and it is expected that a variety of welfare-oriented systems such as medical care, education, traffic control and environmental information systems will be established. Another direction is to create new information-providing media, which have become possible with the progress of communication technologies—particularly picture communication techniques—and which could not previously be realised for both technical and economic reasons.

The start of on-request data and information services, such as PRESTEL in England, ANTIOPE in France and CAPTAINS in Japan, is a new and great revolution for the mass-communications media. This type of information service is expected to develop into a system integrating three communication modes—voice, picture (including moving pictures) and code—like the Video Response System currently being experimented with in Japan. In this system computers play important and complex roles in the overall control, the retrieval of information, and the formation of pictures.

The second area of interest for computer complexing concerns structural reorganisations involving centralisation and decentralisation. Along with the rapid improvement in performance and reliability of the latest generations of computers, the development and mass production of units based on LSI are inducing fundamental innovations in computer utilisation methods. One of these is the partial revision of the concept of scale merit (Grosch's law) leading to a shift from centralised systems based on giant computers to decentralised alternatives. An example is the establishment of the so-called bus-bar system computer network.

More important changes, however, are due to computer miniaturisation. Minicomputers which have been utilised in factory process control are now being substantially further reduced in size, making possible their incorporation as "intelligent parts" in various types of instruments and systems. These microcomputers have already been used as built-in components not only for production processes but also in automobiles and household appliances, so as to make all the system functions "intelligent".

This itself comprises a transfiguration of the computer system, and is contributing to rapid exploitation of new fields of application through the establishment of extremely decentralised systems.

The so-called microcomputer boom currently occurring in the United States and Japan is evidence of a trend towards devices which can be easily used and "enjoyed" by amateurs including school-children. Allowing such amateurs to participate in this way provides important clues concerning probable future developments.

Characteristics of Computerisation in Japan

Early developments and competition

Table 10.1 shows the numbers of units and investments for computers in countries where these investments are high. Table 10.2 shows the ratio of domestically produced to imported computers of various sizes in Japan.

It is evident from these tables that Japanese use of computers is very substantial. Furthermore, this use is increasing at surprising speed, the growth rate over the past few years being as high as 5,000 units/year.

The first fully-fledged operational computers in Japan were introduced by the Meteorological Agency's Forecast Department, with an IBM Model 704 EDPM (Electronic Data Processing Machine) and by Onoda Cement Co., Ltd. with a UFC (UNIVAC File Computer) both in 1959. With these as the starting point, later models of US computers were introduced in rapid succession by many of the larger Japanese enterprises.

In parallel with these expanding computer imports from overseas, very substantial efforts and investments were devoted to the development of domestic computer technology, quickly establishing a solid basis for home manufacture. Principal events are listed below in chronological order.

1954	—	Relay-based FACOM100 completed.
	—	Parametron invented.
1956	—	Vacuum tube-based practical computer FUJIC completed.
	—	Transistor-based ETL-MARK III completed. (The earliest development of a transistorised computer with digital stored-program system)
1957	—	Parametron-based MUSASHINO-1 completed.
	—	Esaki diode invented.
1958 to 1960	—	Six computer manufacturers enter market.

The efforts of Japanese universities, governmental research institutes and manufacturers to develop domestic computer technology and production capacity, have from the earliest stages contributed to making Japan the only other country, besides the US, whose share of domestically-produced computers is now greater than 50%. Moreover, this position has been achieved in the face of severe competition, not only from foreign manufacturers, but particularly among several domestic producers. In contrast to

West European countries, in fact, where one firm prevails in one or a few countries through successive mergers, it is interesting that the principle of competition dominates this field in Japan.

This competition has however been linked with another feature, peculiar to the Japanese computer scene, which is that competing firms participate and cooperate in joint development projects under the leadership of the Nippon Telegraph & Telephone Corporation, or the Ministry of International Trade and Industry, in cases where decentralised efforts for research and development could prove extremely costly yet nevertheless ineffectual. An example is the case of "super LS1" development.

MIS boom

One of the most powerful (albeit transitory) factors responsible for the accelerated development of computerisation in Japan was the MIS (management information system) boom in the late 1960s. The term MIS was brought home by an inspection party which visited the US in autumn 1967 to study the development and uses of large-scale computers for business purposes. The party, which comprised top managers from business circles, studied the most advanced management uses of computers in the US at that time, and reported that Japan should develop and adopt the "total system" (that is, MIS) including its use for top management decision-making.

This is an example of the exaggerated sense of urgency over computer utilisation which swept Japan at that time, producing a crisis feeling of lagging behind Europe and America in the adoption of novel techniques. It was encouraged by the contemporary trend towards increasing "abstraction" in every facet of life. MIS became a magic formula for success, resulting in an "MIS boom" in various fields and strata. In fact, of course, the MIS concept was soon found to be unrealistic in view of computer capabilities at that time and computerised top management decision-making even today is a somewhat utopian concept.

Nevertheless, it cannot be denied that the MIS boom aroused

intense interest in computers among many leaders in business and government circles, and provided the motive force for the sophisticated utilisation of computers in Japan.

Automation and Japanese-style Industrial Management

Industrialisation and technological innovation have been absorbed by Japan over a far shorter period of time than was taken in European countries. Successful adaptation to the new situation has therefore inevitably followed a different pattern to that in other advancing nations.

First of all, very large population migrations occurred from rural to urban communites, comparable in scale but far more rapid than those which earlier affected the Germanic races. For all those who thus left their homeland, a new sense of home was provided by the industrial enterprises and various other organisations of which they became part.

This deliberate and careful cultivation of corporate citizens' communities is one of the major features of Japanese-style management: it means that employees of Japanese enterprises, with situations based on life employment, mostly find spiritual satisfaction in their work, and willingly devote their lives to their jobs.

For its part, management not only makes every effort to use the abilities of each employee to the best advantage of the business, but also assumes responsibility for the employees' extra-mural ("spiritual") growth and satisfaction.

To view the situation in another perspective, while in the European democracies the big enterprise system developed around industrialisation *after* the establishment of rich regional and citizen communities, advanced industrialisation came to Japan *before* the maturation of a democratic community. For this reason, the Japanese enterprises became responsible for the entire level of life under the life employment system, including welfare arrangements provided by the regional community. That is, the role of Japanese enterprises may be appreciated as a unique social development in adapting to a unique social environment.

It is true, however, that Japanese youth today has less sense of devotion than preceding generations to its "guiding enterprise", and frequent conflicts between the egoism of corporations and regional residents are degrading the image of the enterprises.

This unique feature of Japanese-style management—the system of full member participation and life employment—has been of special importance for the acceptance of the automation and computer revolution. It is axiomatic that automation of the production line and computerisation of business management are designed to save labor and reduce costs, improving service by shortening working time and increasing productivity. Therefore, rapid introduction of these innovations must often induce serious labor problems such as unemployment and job changing. These problems were experienced in Japan as in other advanced countries, but in Japan they were accommodated comparatively smoothly, as have been most of the ensuing changes.

One of the reasons for this achievement is said to be the full member participation system in Japanese-style management. For instance, in advanced automated plants such as automobile factories, the production line involves not only the basic ideas of design engineers but also those of field workers, indicating much more flexibility in the man-machine system than that in, for example, Charlie Chaplin's "Modern Times". In some cases, the enterprise is operated with greater regard for full member participation than for immediate efficiency: this additional "mental input", combined with the various other inputs required for business activities (men, materials and money) may well promote subsequent and unexpected improvements in productivity.

On the other hand, it has often been pointed out that Japanese-style management, based on life employment and strict seniority rules, must restrict the mobility of personnel and result in a rigid system which is not adaptable to change. However it may be observed that, while there is little mobility between enterprises, the life employment system ensures high mobility and flexibility within an enterprise (or group of enterprises under common direction). Thus, in spite of successive and extensive changes such as development of new products, modification of production facilities, automation and computerisation, in most cases the resulting personnel problems have been resolved by job changing, assisted by the goodwill of the employees (who plan their jobs and lives on a long-term basis) and by adequate training courses provided by the enterprise.

The attitude of management in these enterprises and organisations is a product of Japan's long history and traditional civilisation, making the introduction of modern technology such as computerisation necessarily subject to social accommodation. This is surely a very interesting example of "technology transfer" within the framework of a long-established and carefully maintained civilisation.

Most modern technologies originated in Europe and America, and were alien to Japan. However, as these technologies settled and matured in the Japanese soil, they began to assume a characteristic national identity. Like many other industrial

TABLE 10.1
Computer installations in major nations
*(As of December 1976)**

Countries	Number of Units	Investments (Million Dollars)
US	61,120	37,287
Japan	20,124	9,622
FRG	10,660	6,527
France	9,186	4,768
UK	8,543	4,794
Italy	4,491	2,119
Canada	3,372	2,301
USSR and East European Countries	15,299	5,494

*Data derived from "Computer White Paper 1977". The statistics do not include minicomputers and other smaller units. Data for Japan from March 1977 Report of JECC Survey Committee.

technologies, automation and the computer are gradually becoming "Japanised".

TABLE 10.2
*Investment share of domestic computers in Japan (As of March 1977)**

	Domestic	Imported
Large	48.8%	51.2%
Medium	71.5	28.5
Compact	73.2	26.8
Subcompact	59.2	40.8
Total	57.4	42.6

*Data obtained from "Computer White Paper 1978". The computer size is graded on the basis of price as specified below.

Large............	250,000,000 Yen or more
Medium	40,000,000 to 250,000,000 Yen
Compact	10,000,000 to 40,000,000 Yen
Subcompact ...	10,000,000 Yen or less

CHAPTER 11

The Problem of Appropriate Technology

U. PELLEGRINI

An Operational Definition of "Appropriate Technology"

In the present age, technology must be seen as a product of the "knowledge industry" in which the inputs are scientific manpower and financial funds for research and development while the outputs are the results of research, applicable to the problems of the individual and of society. Technology therefore is a powerful means for the growth of economy and for the transformation of society: it has to be considered the basic force of modern civilization.

As a whole, investments that mankind has made in scientific and technical research, during the recent decades, are enormous. The world-wide fund of new scientific and technological knowledge is incredibly large, so that today it is more important to increase the efforts on how to apply the results already known than to further intensify new basic scientific research. But this international fund, even if open in theory to all interested parties, as a matter of fact is at disposal only for those countries that are able to manage it. This is the main reason why, up to now, the exploitation of modern technology has been oriented towards the needs and the ultimate benefit of Advanced Industrial Countries (AIC).

If we agree that the essential problem is not technology itself but the management of technology, then we would realize that there is a tremendous amount of technological know-how to be reoriented towards the different needs of each country, according to the quality and rhythm of growth desired for the society we are considering. Therefore, the problem of "appropriate" technology arises.

Speaking of the Less Developed Countries (LDC) frequently the term "appropriate" is used interchangeably with other words such as "low cost", "intermediate", "soft", or "alternative" technologies. But, while these words represent the economic or the engineering dimension of innovation, the term "appropriate" which tends today to be more popular, also includes the social and cultural dimensions of innovation. We could open a theoretical debate on a commonly accepted definition of the term "appropriate" but as a technician I would prefer to rely on an operational definition of this term. As I will describe in the following paragraph, the technology should be considered "appropriate" when its introduction into a community creates a self-reinforcing process internal to the same community, which supports the growth of the local activities and the development of indigenous capabilities as decided by the community itself.

We can describe a community in terms of people, activities they perform, and environments from which they are surrounded. A general block diagram as in Fig. 11.1, derived from a more detailed analysis made by Dr. Chesnut, shows two feedback loops in the behaviour of the community.

The first loop is connected to activities for which the inputs are the three basic resources of each society: materials, energy and information. The second loop is connected to people whose inputs are conditions, expectations and aspirations. Number and capabilities of people as well as their ideas and actions are the transforming factors of the three basic resources of the first loop. The final outputs of both loops are products, services and new ideas, which together in their turn influence activities and people environments.

This general model helps us to better understand the terms of "growth" and "developments" as they have been defined

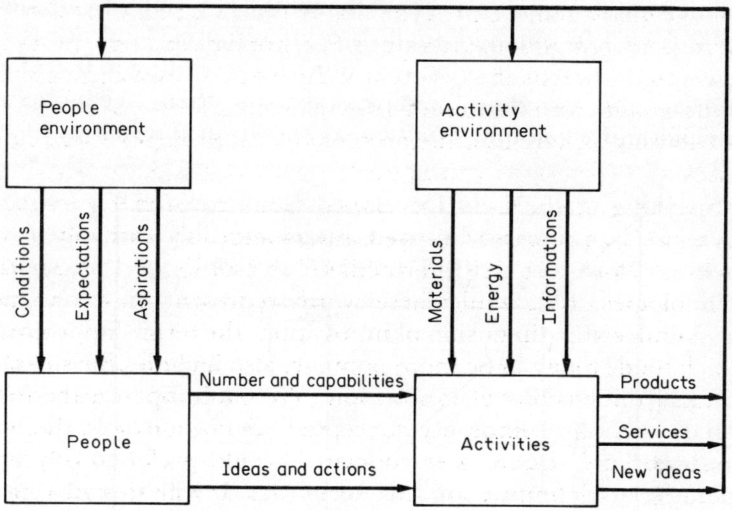

FIG. 11.1. Feedback loops in a system of people, activities and environment.

recently in a stimulating correspondence between M. Cuénod and A. Concheiro, issued in IFAC bulletins.

"Growth" of a country, defined as the increase of its production of material goods and services, can be measured on *quantitative* criteria in terms of products and services by standard and universally-accepted economic indexes: therefore it has to be analyzed with the response of the first loop. "Development", defined as the capacity to solve one's problem using one's resources, is linked not only to tangible resources but even more to the conditions, expectations and aspirations of people, that is to the cultural, social and economic situation of the community: therefore it must be studied including *"qualitative"* criteria which can be found by analyzing the response of the second loop.

Conditions for Success of "Appropriate Technology"

Selection of appropriate technologies for a "quantitative

growth'' and a ''qualitative development'' of a country, implies a previous well formulated analysis of its resources and needs.

Alexander King has classified countries according to their food, material and tangible resources versus their technological, financial and economic situation as shown in the matrix of Fig. 11.2.

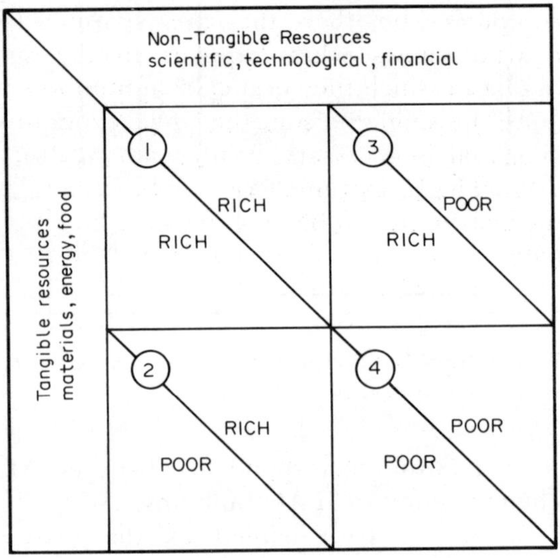

FIG. 11.2. Classification of countries according to their resources situation.

The Western world, the AIC countries in class 1 or 2, are able to develop their economy mainly on the basis of an economic self-reinforcing cycle in which growth creates growth.

This can be seen quite clearly, in the diagram of Fig 11.3, which shows the industry's ability to create and distribute wealth in addition to its ability to produce products. The characteristic of industrial activity to create wealth through the feedback loop shown in Fig. 11.3, often gets little attention or is simply taken for granted. But we have to take care of this basic point from the consideration that the distribution of the wealth already existing within a country has very little and in any case transient effect on the increase of the standard of living of each person. It has no dynamic evolution if not channelled in a feedback loop which increases the global productivity of the country's economy.

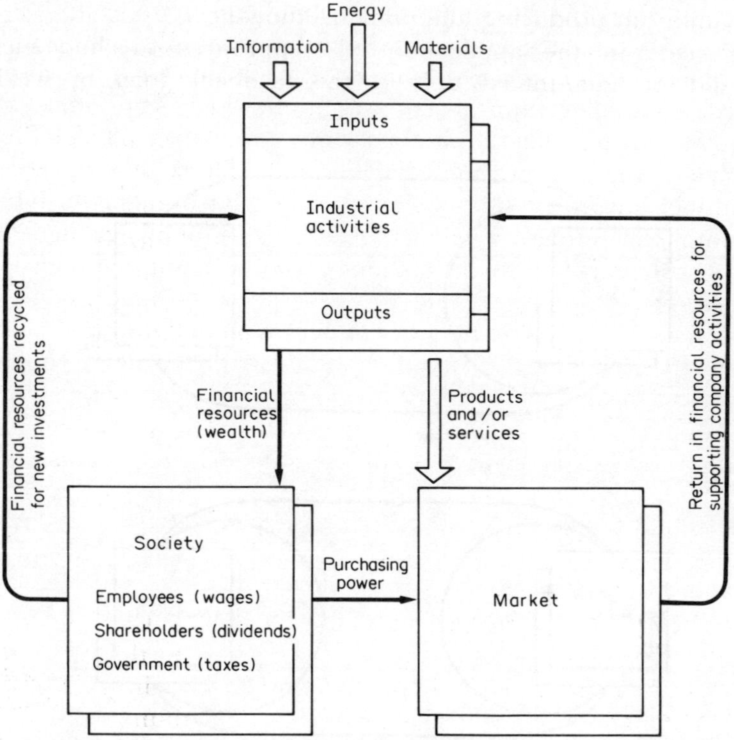

FIG. 11.3. Industrial activities generate wealth as well as products.

This dynamic increase seems to be difficult in the LDCs if the application of a technology initiates a process where a self-reinforcing cycle is settled only between an AIC and the interested LDC, as in Fig. 11.4. (a).

Financial resources are mainly recycled in the AIC with unbalanced economic, political and technological consequences in the LDC: economically it is hard to make interest and capital payment on existing loans from abroad; politically, this sort of feedback loop appears to land mainly in the hands of élites; technologically, it starts only a very slow process of indigenous capability in producing autonomous know-how.

Success in the application of "appropriate technologies" should be guaranteed only if the feedback loop of wealth

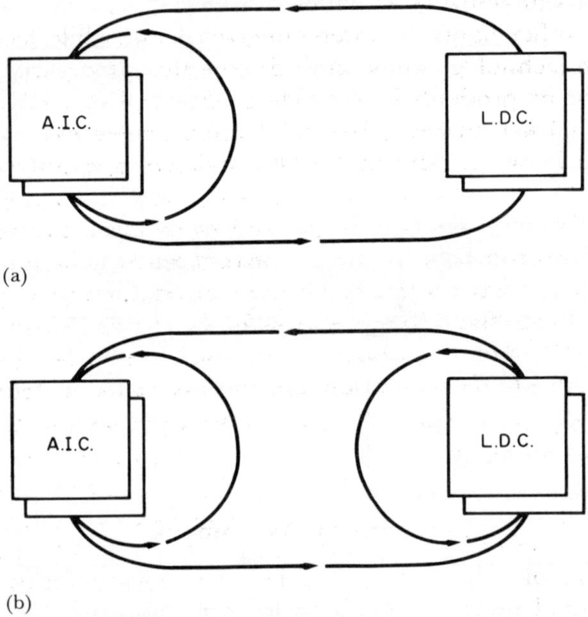

FIG. 11.4.
(a) Feedback loops for NON-APPROPRIATE Technologies.
(b) Feedback loops for APPROPRIATE Technologies.
A.I.C. = Advanced Industrial Countries.
L.D.C. = Less Developed Countries.

generation, described in Fig. 11.3, is settled, not only in the AIC but also within the LDC as in Fig. 11.4 (b).

The latest configuration of technology exploitation has the potential capability of producing social and economic benefits in AICs as well as in LDCs due to the internal self-reinforcing process present in both interested parties, and to the interconnected loop which assures mutual cooperation.

But why should an AIC consider it an advantage to help starting a self-reinforcing loop in a LDC? This difficult and, often, intractable question could be answered looking at the new potential markets created by the increasing incomes originated by the new manufacturing activities in LDCs.

On the other hand the interconnected loop, while feeding the LDC with technology know-how gives to the same party an open outflow for its products in the richer market of the AIC.

So the market function shall reinforce the interconnected feed-back loop between AIC and LDC and will guarantee mutual interests.

When this loop tends to be broken by political procedures of national Governments, or the interaction tends to be forced one-way only, it is hard to maintain the self-reinforcing loop within the LDC due to the different speeds in the dynamic behaviour of technological and economic developments in the interested parties. Indeed, the condition for success is to be found in a systemic approach that maintains in operation simultaneously all the feed-back loops.

Technology Assessment

The task of selecting appropriate technology for satisfying economic and social needs, is to forecast the range of possible effects on people, local environment and transnational cooperation before options are foreclosed by default. This is a multivariable problem regarding a complex system with many components interacting in an interdisciplinary mode. The main components and the range of potential interactions, at very gross level, is shown in Fig. 11.5.

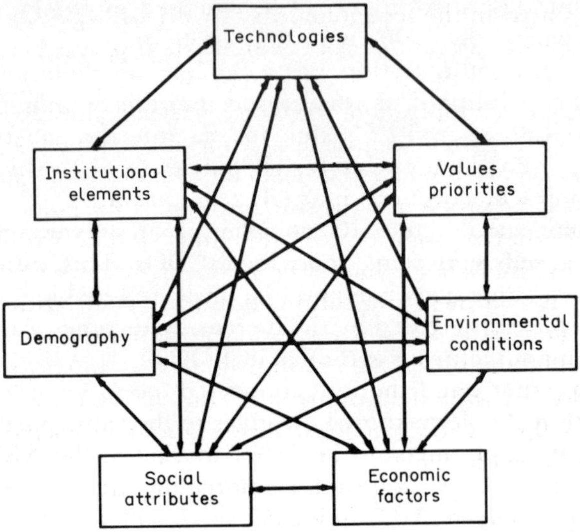

FIG. 11.5. Some gross-level interaction.

If n are the components, the potential interactions are n (n−1); if each interaction has only two values (existence or non-existence), which is an underestimation of the continuous range of values they normally have, we find out that the possible states of the system are

$$2^{n.(n-1)}$$

Therefore, the process of technology assessment, even when the analysis is reduced to a much simpler level because some interactions are not taken into account, is a complex task and needs a formalized procedure to be managed.

In the past years various methodologies have been developed. The different schemes of procedures adopted can always be reduced to three successive phases of analysis (technological, socio-economic, and political) as shown in Fig. 11.6.

Exploring the scenario of alternatives in function of the

objectives stated in the technological assessment, the analysis will result in a cycling iterative process; in the first phase, science and technology point out the available means to perform; in the second phase, economy and sociology indicate how to evaluate the impacts of applied technology; while in the third phase, the choice must be oriented by political considerations. Definition of the technological problem is the starting step, while political choice between alternative solutions, is the final step of this iterative process. The final choice has to be political, because technology must be seen as a means to fulfil community goals. If a nation's overriding goal is to maximize national income, the technologies used should be the most efficient in productivity in relation to the factor "cost" within the country. If, on the contrary national goals stress the creation of jobs, more labour-intensive technologies should be chosen.

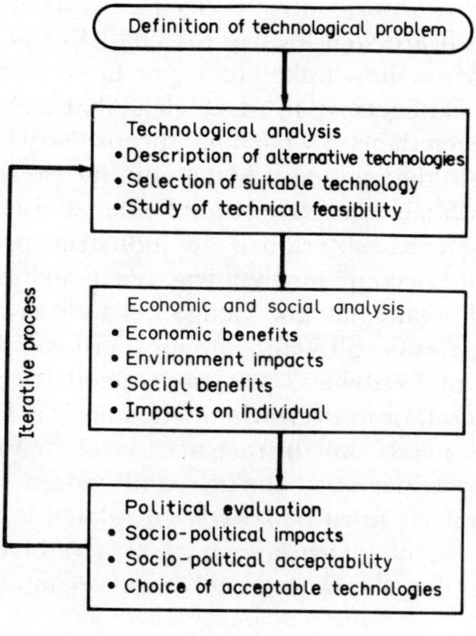

FIG. 11.6. Technological assessment.

Nowadays, owing to the broad spectrum of diversified developments, there is always a plurality of technologies to be applied for a given industrial-economic problem: from capital-intensive to labour-intensive technologies, from centralized to distributed solutions.

Technology Transfer

Technology transfer comes after technology assessment and from an operational point of view is of paramount importance because the efficiency of technology exploitation depends on its implementation.

Transfer is accomplished through many routes and by many means. Selling products manufactured abroad in a local market is a form of commercial transfer by the use of technological goods; information exchanged through books, journals, technical reports, or generally printed papers, is another way; but frequently they are not controlled by local communities. Nevertheless, as we all are well aware, the challenge is the transfer of processes and know-how linked to a specific economic activity. In this respect, it is necessary to make clear that technology seldom involves, and even then only partially, proprietary knowledge to be negotiated with the owners. Much of the present knowledge developed in AICs is normally of public domain as it is embodied in published scientific literature for industrial processes, which nowadays, could become newly attractive in a changing situation of resources. For all that knowledge, as well as for proprietary knowledge of patents still valid, the most critical action is not the mere transfer of hardware, processes, goods or papers. For a knowledge exploitation, even when technology is freely available, a community needs an infrastructure of manpower, skills, innovative capacities to absorb and adapt the imported technology: transfer must be strictly correlated to the acquisition of the know-how by community people themselves that have to participate in the shaping of technical economic and social change.

In the LDCs, industrialization, accompanied by the acquisition of related technologies, should not simply consist of a replication of the structures and policies in an AIC. It is important to ensure that the path chosen respects the cultural heritage and social values of the community concerned.

Therefore investments for education and training play a decisive role. These cannot be limited only to a coordinated technical assistance program for manpower training and for developing managerial capabilities in the specific industrial sectors concerning the transferred technology. It involves medium-long term investments to create a ''humus'' in the community by establishing institutions for education, universities and research centers. In this connection Governments have to define a national science and technology plan which has to be linked to socio-economic planning. Indeed the long-term investments in people will finally determine the success in creating the self-reinforcing process of ''appropriate technologies'' for the development of the country.

Conclusion

We have seen that the problem of defining and implementing the ''appropriate technology'' for growth and development of a country is a multivariable, multicomponents problem that has to be analyzed with a formalized approach. The closed feedback loop aspect is inherent to the evolution of technical, economic and social structures. Therefore the system theory and the control theory are the disciplines that can offer the methodology of problem-solving in this complex, interdisciplinary context.

The mechanism for analyzing highly complicated non-linear dynamic models is something that is well understood by many specialists.

But, till now, too little research has been done in a cooperative effort, putting together systems analysts, economists and sociologists.

Therefore, to handle the problem of ''appropriate technology'' in a world of increasing inter-related varieties, which are a consequence of technology itself, we need to promote interdisciplinary efforts, as is done in the Discoveries Symposium.

Bibliography

1. H. Chesnut, *Influence of Technology on Modern World*, Lecture given at Accademia Nazionale dei Lincei, Roma 1972.
2. F. Hetman, *Society and the Assessment of Technology*, OECD, Paris 1973.
3. S. Beer, *Platform for Change*, John Wiley, London 1975.
4. N. Jéquier, *Appropriate Technology: Problems and Promises*, OECD, Paris 1976.
5. A. King, *Scientific and Technological Imperatives for a sustainable Society*, Conference on Alternatives to growth, The Woodlands, Tex. Oct., 1977.
6. *Pugwash Guidelines for International Scientific Cooperation for Development*, Geneva, December 1978.
7. L. Benton, *Management for the Future*, McGraw-Hill, New York, 1978.

CHAPTER 12

The Impact of Automated Information Processing on Society

H. ZEMANEK

Introduction

Both the Discoveries Project and the subject of this paper are very close to my heart—I consider them as important as the scientific fundamentals of information processing and I have been active in pursuit of a better understanding of the human aspects of information processing for many years.

Perhaps a few words of explanation as to how I entered this area and formed my opinions would be useful. My original professional background is telecommunication engineering, but by around 1950 I already had turned to cybernetics, computers, and the information sciences. With a group of former students of mine we built one of the early European fully-transistorized computers at the University of Technology, Vienna, and later, in the IBM Laboratories in Vienna, we developed with the same group of people what is now known as VDL, the Vienna Definition Language and Method, by which we formally defined the important programming language PL/I—not only the syntax—its grammar—but also its semantics, the meaning of every possible program in this language.

Dealing with language and semantics means dealing with one of the most human chapters of science. There is a common feature of programming and cybernetics: the tension and the gap between the informality of natural systems and the strict formality of logical description, scientific models and digital programs. In cybernetics, the scientific and technological view and description of live structures mean an increase of precision of description, up to the simulation of the structure, for the price of a clear reduction of the functional properties.

In programming, the languages have formal character, but their development from idea to concept and further to the defining document is a transition from intuitive and informal notions to logical mechanisms; either this defining document is extremely voluminous and hard to read or else many details are incompletely defined. And this dichotomy is typical for all information processing activities beyond pure logic and mathematics, and even there one can find many more informal corners than one might expect.

Technological development has reached a density in which a revolutionary technology like information processing must bring turbulence and dangers which should not be allowed to grow arbitrarily. The computer *is* an invention of exceptional character. Apart from its extreme intellectual importance, the computer has the tendency to intrude into everyone's business and increasingly permeate the privacy of the individual. By necessity a fraction of the resulting effects is negative. We experts not only owe the layman an explanation for what has happened—we also owe the remedy or, when it is beyond our power, at least the offer of remedy.

All of this is the result of the very intriguing character of the entity the computer is dealing with: information. The computer has been called a thinking machine, and is the basis for a field called artificial intelligence. But the thinking power and intelligence of the computer are of a specific nature: of an algorithmic nature; and an algorithm is always the final result of a lot of human thinking including some real or only hoped-for perfection. Perfection, however, is the exception in our real world, and so there is a constant need of human interface between the operation of a computer and consequences in the real world. It is this need of human interface which makes the computer a complicated social issue.

Expressed in one sentence, the impact of technology in general is amplification and extension of human power for the price of increased energy consumption and a loss of natural completeness. Information processing is amplification of speed and precision of that part of mental power which consists of logical derivation; information processing requires less and less energy and can help to save and reduce energy consumption. But information processing may lead much further away from natural completeness than conventional technology.

The impact of information processing can naturally be classified by technical, sociological and philosophical impact, and so I will discuss it, adding a short summary of the impact on developing nations.

A Short Story

It is very inviting to see computer perfection as a guarantee of success. But perfection exists only under certain conditions. Let me illustrate this by a little story, a true story that I heard from one of the persons present. It was in the early days of computers, not very long after World War II. Philips and IBM had decided to discuss certain things of common interest and a small IBM delegation stayed for some time at the Philips Laboratories at Eindhoven. Hotels at that time had not entered the boom era of business trips and social tourism; the delegation was housed in a modest, old-fashioned place.

Some members of the delegation wanted 3-minute eggs for breakfast, but were always served with hard-boiled eggs. Most of the party accepted this as fate, but one member was particularly keen to get a 3-minute egg and tried every day to make the cook comply with his wish. The delegation was amused at his attempts and, as the end of their stay approached, started making bets as to whether or not he would be successful. The last morning came, the last opportunity—would he make it or not? He called the cook and carefully rehearsed the algorithm for a 3-minute egg across all Dutch-English language difficulties. "Have the water boiling, then put the eggs into it, and after no more than three minutes take them out and serve them. Do you understand?" "Yes", said the cook and disappeared into the kitchen. Full of tension the delegation waited for the result. It was a hard-boiled egg. After a lot of laughter they investigated this defeat and found out that the hotel received every morning, for some unknown reason, a consignment of cold, hard-boiled eggs.

The most obvious and absolutely correct algorithm may be wrong and even harmful if it works under incorrect assumptions. It is essential to understand that the algorithm, the computer, the information processing system, are only part of the practical application, and that the environment, the assumptions before and after processing, the collection, interpretation and application of data as well as the reactions of the people and the consequences for the people close to the information processing system, or far away from it, make up the other, equally important part of the application. Information processing is not limited to the section between keyboard and printer, it is functionally embedded in a live environment, in society and its sub-groups.

Automated Information Processing

Before discussing its impact on society it is reasonable to discuss the nature of the subject, automated information processing. Information processing is a technology, regardless of whether we are dealing with hardware or software—which means that not only circuit production but also programming is an engineering activity. Mathematics and logic are fundamental but auxiliary sciences for programming, and program design, implementation, maintenance and application must be carried out in an engineering spirit.

Technology is the application of scientific and mathematical knowledge for the solution of practical problems, at the same time respecting the principles of economy. Ideally it is expected that technology achieves things with less effort and fewer costs, faster, stronger and with more precision. A critical consideration of technology must not forget how well those aims have indeed been reached in many fields. Only as a second step one will have to consider where increased effort, higher costs and longer delays have cast a shadow on this ideal picture. Also, it does not make much sense to formulate wishful requirements that go beyond the economic foundations of a certain field. Who is ready to pay for an achievement is a basic question. Beauty, safety and quality of life have their price, protection of groups or individuals does not come free of charge, demands are bound to obligations.

Automatic technology is in no way a modern idea. Just think of the mousetrap, which is a true automation several thousand years old and which has all the basic features of automatic technology like power supply, information input, processor and effector, and which works, last but not least, without human supervision. Another invention several hundred years old is the semi-stable multivibrator which we trigger in the water closet. It has such a beautifully simple architecture that hardly anyone cares to know what its functional diagram looks like. Another feedback mechanism was added to the steam engine: the governor, first computed by Maxwell in 1868. What electronic engineering has added is simplicity of realization, and what the computer has added is the mastering of almost arbitrarily complicated and sophisticated programs. As a consequence of the fantastic improvements in semiconductor technology—modern technology makes automation not only available to big companies and institutions, but to the general public and to every household at such low cost that the average household in a developed country will soon have as many processors as it has electromotors today—please count the motors in your home!

Telecommunication technology is the basis of information processing technology. It is perfectly correct to describe the world telephone dialling system as a distributed computer. And the digital technology of information processing is a heritage of radar technology which also is a kind of computering equipment, although of analogous nature. All those roots are worthwhile mentioning not only in order to understand the essence of information processing, but also because the impact on society will have a lot of similarities.

Let me mention two examples of this impact which are more or less repeated by the computer. Telephone and television connect people and separate them at the same time. At least in principle, one can talk today from any telephone to any other telephone in the world; things can be arranged independent of geographic distances, people on a trip are never separated from their office or their home unless they wish to be. And yet there is a separation effect: what once was managed by a visit during which one got a feeling for the personal situation of the partner and had an opportunity of dealing with many subsidiary problems and side-effects, today is organized by a telephone call, interrupting, maybe, the work or a conversation of the partner. Automatic telephone calls do not offer priorities for people or subjects, but the telephone has a kind of automatic priority over the non-electronic atmosphere in which the telephone is located. Computer terminals can be equipped with storage capabilities and information processing systems can be programmed with priority structures—but the impersonal organization power of computerized systems goes far beyond the psychological force of the telephone buzzer.

Radar technology is an example of information collection by electronic means—unexpected, unwanted and unnoticed by potential victims, in war, road traffic and elsewhere. What is a useful tool for one application, air traffic control for instance—as long as the officers are not on strike—may intrude into the private sphere in another application, and these contradictory effects lie close together. Information collection by means of computers is, of course, more effective for many orders of magnitude.

Information technology is connected to several scientific and technical fields, and they are in a process of unification under the overwhelming power of the computer. Electronics, including electronic switching functions, transmission and storage of digital information are the hardware basis on which many levels of programs, programming systems and programming languages are built up.

There is information science which one day is expected to cover all sorts of information needs and information services, from the newspaper and news broadcast to the library and computer-based information systems for documentation and information retrieval. Here, a lot of theoretical work remains to be done and, much to the regret of classic information-processing, the scientific foundations for those services proceed very slowly.

Therefore, information technology and computer technology to a large degree are identical fields and we can turn without neglecting too much to information processing in a more restricted sense, to the magical tool of the twentieth century—to the computer.

The Computer

The classic computer is best described as a character storage and replacement device: controlled by strictly logical rules, one set or sequence of characters is replaced by another one; the process can increase or decrease the total amount of characters. It is easy to show that a logical device of this kind can transform, select and destroy information, but never generate information. If it does not generate more characters than are fed in, it produces redundancy and not information.

It might be useful to dwell a little longer upon the notion of information. The first aspect to be taken up here is the *formal* nature of information, indicated already by the common root of the two words, namely form: a formless pattern gives no information although it can be described point by point—thus requiring a lot of information. But for our purpose we need not specify the distinction any closer because the digital computer handles exclusively characters and remains entirely within the universe of formal relationships. This is less a restriction than it might seem. Because the human language, spoken or written, remains also under the formal principle: a spoken sound and a written letter are tokens out of a very finite set, of an alphabet of sounds and letters between 20 and 40 members. Of course, there is further information, e.g. in the melody of speech; such information can be sensed by the human receiver—it is normally lost before information processing. Man and computer can transform any kind of informal information into formal description.

Compared to the human brain, the computer imposes a more severe restriction; the computer operates *strictly by logical rules.* And these rules must precede any processing—very much in contrast to human information processing where logical rules are the final result of long, complicated and error-suffering thinking processes. The best representative of operations according to logical rules is the algorithm; non-algorithmic information processing is possible but risky, and there are only rare occasions to proceed in a non-algorithmic way, for instance in an exploring mode. The algorithm is a prescription which leads in a finite number of steps for a finite or infinite number of possible input data automatically to the correct result. Think of your way to multiply an n-digit number with an m-digit number—a typical example for numerical calculation*. The 3-minute egg algorithm may not resist closer investigation: it may turn out that it gives an incomplete description of the process wanted; it might, however, also happen that it is precise enough: if either the egg-cooking system is well enough defined to be satisfied with the given information—in fact it could be sufficient to have the name "boiled egg" and the value "3 minutes" as an input—or because the resulting ambiguities do not matter.

This comparison of a numerical algorithm and a processing rule out of our daily life shows the wide range of computer applications and the variety of linguistic means to state problems and to describe ways of solution and results.

One could insert sections about language questions at any point in this paper and they would never be enough, even if the same point is repeated in different variations. The multiplication of two numbers is a formal problem which does not require any explanation or interpretation beyond elementary mathematics. The preparation of eggs may seem to include much less complications because it is something one can learn fast. But in fact one

*This multiplication algorithm belongs to the set of algorithms given by al-Khorezmi, the Arab mathematician (780–850), from whom the name algorithm is derived. He expresses his algorithms in natural language—formal notation is an achievement dating to the Renaissance (to what extent the ancient Greeks were forerunners is a matter of discussion).

may say that the simpler the problem the more difficult its linguistic aspect. Everyone knows what is meant by words like "egg" and "cooking", and yet only very few people would be able to give sufficiently precise definitions of the two notions. This is true for all the simple notions: can you define notions like street, veal, government officer, client, transfer, keynote address? The arrangement of letters denoting these notions can easily be fed into the computer—but there they are separated from their meaning and who knows what happens in the subsequent processing of a text in which they occur, or of numbers related to them? Sometimes the single person, the single programmer may know everything that is necessary. With team work it becomes already more difficult. In extreme cases, nobody knows.

We will come back to the problem of meaning and its conservation in information processing. At this point we will only say that the computer has no notion of meaning, it deals exclusively with formal structures and expressions. And these formal structures and expressions have no meaning whatsoever unless they are used in relation to an informal environment. Indeed, the philosopher Ludwig Wittgenstein, in one of his radical statements, has called logic a tautology. In itself, the computer is a tautologic device, the meaning is outside, in the human brain.

As to its formality, the computer is as radical as Wittgenstein; all computer processing is transformation of thoroughly formal structures and expressions, down to unequivocal bits—whether hardware or software: what cannot be analyzed down to logically arranged bits is either noise or misfunction. The miracle is that we master the basically non-digital electronics so well that the computer is a digital entity even in the most philosophical sense. Of course, one can simulate non-digital variables and compounds, but wherever one looks close enough the true digital character becomes apparent. The computer, however, is always connected to an informal, more or less natural environment, and finally to the people who program and operate the information processing system. And even programmers are not formal, rational, digital thinkers—neither as individuals nor as professionals—whoever has had to manage programmers will know and agree.

If information processing as a technology and profession has already so many and important human aspects, its impact on society must correspondingly have many and important peculiarities. There is no non-trivial engineering innovation without an impact on society. Many political events of the last hundred years were, directly or indirectly, triggered by the progress of technology, although not as many as certain mechanistical political theories would like to have it. In any case, it might be argued that the computer is not exceptional and that the present emphasis on the social consequences of automated information processing is merely a fashion which will not last long. Such an argument is under-estimating the power of information processing and a misjudgement of the motivation behind the present, highly emotional evaluation of technological developments.

The System

It is information which makes the whole more than its parts: by means of information sensors, processors and effectors the parts can be integrated into a system. The living organism is the fundamental example for a system; survival would be impossible without information processing.

The basis of life is the cell, the subsystem which has, in principle, all the faculties of the whole system, the complete organism: it absorbs and releases material and energy, it reacts to pressure and light, it is able to reproduce itself and to organize higher structures. It is remarkable that nature chose a digital code for the description and automatic reproduction of each cell and any living organism from the most simple one up to the human body.

It is, therefore, not astonishing that systems theory originated twice, once in biology and once in telecommunication technology. Information processing systems do not have all the universal properties of the living organism, but they come closer to it than any other technological entity. And they can integrate all kinds of ensembles to self-controlled automatic systems, matching even changing environments. The same is, by the way, true for feedback theory.

It seems very remarkable that facilities and requirements in certain developments are so much interlaced that it can become very difficult to distinguish which impacts which. Do we need systems because we can now organize them or are we able to organize systems because we need them?

Until very recently most technological innovation could be seen as a singular solution, improvement and phenomenon. But now we have reached such a density of excellent solutions and effective improvements that they hamper each other and produce chaos. The outstanding example for this effect is road traffic. Never before has mankind had such excellent roads, so many traffic signals and such powerful, fast and well-designed vehicles. But the sum of all these excellent solutions and effective improvements are king-size traffic jams. We are much better in singular solutions than in systems design, and not only in traffic—this is true throughout technology and there is no risk in predicting that information processing will see itself soon in the same trouble. Even if the single, centralized big computer will remain a tool for order, the mushrooming networks and mini-processors will achieve a chaos of equal or even bigger size.

We would not be in the field of information processing if the proper solution was not available. It is called architectural design. It is not astonishing that we hear and speak more and more of it, that structural programming is widely accepted and that everyone likes top-down design. We have to go this way, although we are still far from really mastering system architecture.

The computer is in manifold relation to architecture. It should have a good hardware and software architecture and it can be used to achieve good architecture in the design process for any kind of systems, whether circuits, system or application programs, and of course also in the classic building architecture.

Architectural design of computers has been introduced on a large scale by the IBM System/360, the first wide-range family of computers ever built; 360 stands for the full circle. The whole family of processors, memories and other equipment was governed by the same design principles, the members of the family distinguished only by increasing or decreasing values of certain parameters. Since System/360, architectural design has been a keyword. On the hardware side it is inherently imposed by automatic production of the miniaturized circuitry via almost fully automatic production lines for the chips. Here, the computer becomes a tool of architecture: computer-assisted design is a necessity for such automatic production. Proposals for hardware which do not fit into the design programs and into the architecure of the production line have no chance to be realized, and this alone imposes architectural discipline on newly designed systems. On the software side, architecture, it is true, could be less of a fashion and more of a discipline, but the advance in systematics is clearly visible.

The progress in architectural virtues is badly needed. Because the architecture of the information processing system evidently influences strongly the architecture of all applications and consequently of all systems dominated or even only related to computers, much more than applications can feed back on computer system architecture.

A big computing center with its multitude of processors, storage units and peripheral equipment is a system which raises many problems of efficiency and cost/performance, and it must be admitted that there is not yet much of a useful system theory to support its design or configuration. The advance of digital communications and the growing need for distributed information processing and the reduction of line cost makes it possible to distribute input, processing, storage and output over cities, countries, continents and the whole globe. If the computer started as a centralizing power, it has meanwhile developed into a tool for decentralized centralization or centralized decentralization. If properly conceived and designed, automatic information processing can provide a maximum of information, decision liberty and human choice to peripheral subcenters without reducing the efficiency of central policy and overall success. Wherever a system suffers from bureaucratic inflexibility, slowness and inefficiency, the cure can be well-designed distributed computer assistance which re-establishes the flexibility, speed and success possible in a small structure. I am not saying it is easy to achieve that, I am saying that there is a chance and that much emphasis must and will be laid on development in this direction.

The Technical Impact

Almost inadvertently, we have already begun to discuss the impact of automated information processing: its first and most important power is concentration and decentralization, offering flexibility, reaction speed and a clear picture of the momentary situation—the basis for subsequent action.

But before the impact is discussed in a more systematic way, it seems reasonable to consider what the computer contributes to the world of today, to science, to technology, to industry and economy, to administration and government.

First of all, it has become possible to carry out numerical and logical elaborations of considerable volume; the solution of systems of equations, for instance, including an amount of elementary operations of which previous centuries could not even dream. It is true that mathematicians as well as nature can always provide problems exceeding any existing computer facility. Just take weather forecasting as an example; here it becomes once more obvious that computing alone is not enough—what is needed in addition is input information in considerable quantity for which numerous sensors, not only on the surface but quite high up in the atmosphere would be required. Voluminous computations meet a lot of limitations, but the achievement has been impressive for the last two or three decades.

The same is true for commercial and statistical data flow. Accounting, inventory control, payroll problems of any size are standard applications, and with the distributed computer, with data networks, it is possible to run live data-bases for all kinds of purposes. Rather than to try and enumerate or list all possible subjects, we will pick a few typical examples and demonstrate the general effect by the special case.

An important special case of data flow and data-base is electronic fund transfer systems. They are a further step in the abstraction of property and financial power, from goods to gold and silver, to bank notes and cheques, to credit cards and finally to screen-operated money transfer. Extreme mobility of money is connected with instant reporting. It is only one step further to consider general message transfer, the combination of telecommunications and computer systems to general information systems. This combination outdates all classical means of information and money handling.

An excellent example to study the impact of electronics and information processing is the postal system. Where mail and telephone are under a common administration, the loss on mail due to transition from letter to call can, at least for some time, be compensated by the earnings on the telephone. But in the long run this cannot help. The mail service is bound to degenerate and to become more expensive at the same time. The telephone has replaced a high percentage of letter writing and all really urgent news goes by telephone with or without written confirmation later on. True first-class mail, in other words, has disappeared. It is estimated that in North America 60% of the remainder of the present mail is connected with payment procedures; as soon as information processing provides the necessary facilities on a large basis, it will swallow this part of mail traffic, and not only this percentage will disappear from the total volume of mail—the required terminals can equally well handle most of the written messages so that whoever owns a terminal will reduce the number of mailings. The increased number of installed terminals will decrease their price, will facilitate mass production and so further decrease the price of equipment and service. Moreover it will attract other fields like publicity and advertising and so further reduce the number of mailed items. The postal administration will in no way be able to compensate for those losses—at best it can remain in business by providing the new competing services. Since the costs of the mailing system are to a considerable fraction independent of the traffic volume, the price per action, say the cost of a standard letter, must go up. Even worse, in their attempt to keep up with the competing services, the mail and parcel service must invest into automation and then try to get a return on their investment.

Like every other case of technological innovation, the advanced solution cannot cover the full spectrum of the old-fashioned system. The artistically designed postage stamp, for instance, is becoming obsolete, and many things that were possible in the classic mail system are bound to disappear or are in danger of disappearing, will be reduced in frequency or become luxury items—the world of letter sending and receiving will soon undergo drastic changes.

It is customary for defenders of progress to depreciate or ridicule the nostalgic feelings of the average man, his longing for bygone eras. Certainly, a part of this regret is a consequence of the fact that we fortunately remember agreeable things more easily than unpleasant ones. But another part of the longing for bygone times comes from the one-sided character of advancing technology which eliminates natural possibilities by the restrictions of the new system. You just have to imagine the mail being totally replaced by a computer terminal system with enough capacity and intelligence locally and in its channels: any message will arrive at its destination in less than half an hour. There, it is either read and answered immediately or stored until it is used; push a button and a hard copy of the message comes out of the terminal.

But you cannot attach a photograph, a pressed flower or a little specimen of something to the electronic message; you can only read from the handwriting of your partner about his personality and his mood. Have a look at your mail items for a couple of days or weeks and imagine their electronic and computerized replacement. Consider that the poet, the scientist, the philosopher of the fully computerized age will leave to posterity tapes and disks instead of manuscripts and notes. Then you will know a lot about the present and future impact of information processing on our life and our society.

The Social Impact

It is the purpose of technology either

—to perform conventional tasks with less effort and at reduced costs, or
—to perform new tasks which had been physically or economically impossible without the new technology.

In the first case, technology obviously is a job-killer—it reduces the manpower required and it tries to save money—where else in our times than on the salary sector?

In the second case, technology is a job creator, and nobody can question that technology is the only way to keep the present population of a continent like Europe fed and working. A computer-like example for job-killing and creating is the typewriter. It certainly was a job-killer for scribes, but their number was in no way comparable with the number of secretaries and typists of today. We would laugh heartily today if we found a statement from a sociological pessimist who had just seen the first typewriter in which he objects to the job-killing dangers of the new invention.

The analogy for the conventional computer as well as for the mini-computer is obvious. How much more than the typewriter does the computer initiate an avalanche effect: one computer put into operation prepares the need for further computers. Information processing technology is a job-creating invention. But this is not the full story.

A man whose qualification was the production of calligraphic documents will feel little consolation when he hears of an increase in jobs using a typewriter. The problem is transition and ageing. As long as technology develops at such a speed that obsolete jobs die at the speed at which people retire, a modest educational opportunity will be sufficient to maintain the professional balance. In our days some job requirements change faster than our best educational measures can compensate for. This is a social problem and it is reasonable to sit down with all those concerned and responsible, and discuss what can be done.

There is a second effect which aggravates the first, and that is the dehumanization which abstraction and labor division bring to the working place. A medieval shoemaker like Hans Sachs produced a pair of shoes out of raw materials he prepared either himself or bought from people he knew very well, and he sold the majority of his shoes to people he knew very well.

The distance between suppliers, producers and users in modern technology is steadily increasing and automatic information processing does more to separate them further than to bring them together again. The single working place tends to be a narrow slot through which it is almost impossible to see the complete product. In a big information processing system no single person knows all the details.

All of this has very many facets; some features are a necessity, others occur stochastically because nobody cares.

General principles of well-ordered design and disciplined architecture, human rather than logic and algorithmic criteria, give much hope.

Information processing will again and again make clear that perfection is an idealistic concept, fine for models of what we are doing and investigating, and a necessity for logical and mathematical constructions, but hopeless for the live world in which human imperfection is principle and challenge, obstacle and consolation. I cannot think of a better impact on society.

Furthermore, as an information machine of high speed, the computer has a further, very comforting property: it makes us see dangers and challenges far ahead of their real occurrence; so they are observed, described, modelled, published and discussed when there is still time to meet them. And they can be met, as the subject of privacy, well-known particularly in this country, proves.

It is natural for me to add another confirmation that challenges and dangers in information processing are recognized and that proper reactions are initiated, and this in the field of professional societies, because I had myself an active role in them and in the specific development of paying attention to the social impact of information processing. A similar story could be told about

IFAC, the International Federation of Automatic Control, with whom we had close cooperation, but I will restrict myself to a few details on IFIP, the International Federation for Information Processing.

In April 1974, together with some friends in the Austrian trade unions and my colleagues in IFIP and in the Austrian Computer Society, we organized the first IFIP Conference on Human Choice and Computers, where—for the first time in the history of information processing—computer and social scientists came together with representatives of employer and employee organizations to discuss common problems around the computer. It was very impressive to observe their struggle for a common language and for mutual understanding of very diversified views on a young technology. In August 1974, IFIP held its 6th World Congress in Stockholm, giving as always an overview of the state of the art of information processing and including extensive reports on the Vienna conference. Not very much later, IFIP created, in spite of considerable internal resistance, a committee on social questions in information processing. Early in 1979 IFIP had a conference in Budapest on the Socio-Technical Aspects of Computerization, which was a forum for the views of the Socialist countries on the subject, and in June there was the Second Vienna Conference on Human Choice and Computers. There is no doubt that the professional world of information processing has now recognized the social importance and makes many attempts to understand the good and the bad consequences of the computer for society. I could easily present a list of almost a dozen further conferences on its social impact, in many countries and continents, for developed and developing countries.

The Fourth Discoveries Symposium, organized by the Swedish Academy of Engineering Sciences and sponsored by the Honda Foundation, two bodies representing two countries important for technological progress, had several special features, and the invitation to give the opening address was a particular challenge to me to try and summarize the situation as I saw it. The result may have had very individual facets, but since the symposium treated all questions during one week, there was ample opportunity to correct and improve the contents of this address. I saw my main task in exciting fruitful discussions, and the present paper also is presented in this sense.

The Impact on Developing Countries

For developing countries the technological, social and philosophical impact is essentially the same, only it may be much stronger, more helpful and more risky; and as soon as computers have started to multiply, the consequences may develop much faster. This acceleration might not be as good as it seems.

For a short time of my life, during World War II, I was delegated to the German Acceleration Office—*Reichsbeschleunigungsstelle.* The rocket engineers had realized that the starting accelerations could be catastrophic for the electronic hardware in their vehicles unless all effects were carefully studied and tested. The same must be true for the human brain in cultural and technological accelerations, but many, even leading people do not seem to realize that. I have encountered naive views as if the problem were simply one of terminal attachment to the data-bases of Western know-how—and that thereby the leap from the Stone Age to the Space Age would be accomplished.

My personal view is that the computer is not a natural friend of developing nations and that a big part of the present acceleration attempts are more than questionable. Why should the developing nations hurry to reach the spot where developed nations are so unhappy to be?

The computer can be made, I think, into a very useful tool for developing countries—if applied with care and patience, with understanding and discipline. Certain information can go ahead of development and help it, most information—meaningful information—can, however, only be a consequence of development.

In other words: nothing is more difficult than to help, and it is more difficult to help by means of information processing. The gain in technological facilities may easily be followed by severe damage to society and mentality.

Information processing for developing countries has for years been a matter of intensive studies in the UN family, in the international professional federations and the Intergovernmental Bureau for Informatics. Many actions have been studied, but it does not seem that the overall success is really convincing.

The Philosophical Impact

The technological and social impact of information processing results in changes in products and working methods, in jobs and in job-satisfaction. Proper technology and proper management, cooperation between employers' and employees' organizations based on the reality of the working situation and on insights gained by computer and social scientists will find a way which might not totally avoid certain hardships, but which, in the long run, will find acceptable solutions at an acceptable price.

But there is also an impact beyond scientific, technological and sociological details, the impact on and of mentality and philosophy of all the people working in information processing or being influenced by it.

A discussion of this impact could cover even more items and aspects than all the subjects which I have already discussed. Again I will select only a part in order to characterize the set, and I will do this in two chapters. First I will return to the notion of meaning, which I have left insufficiently treated, and then I want to consider possible influences on the mentality of the "average man".

The scientific investigation of language has shown that there is no lowest boundary for the concept of language; the theory covers every level from extremely primitive sign systems up to the most subtle possibilities of human language; and, of course, all of it can appear in the computer context.

The scientific theory of language distinguishes between three levels of increasing importance: syntax, semantics and pragmatics.

Syntax is the study of symbol and word arrangement, the rules for well-formed expressions, correct spelling and correct computing. The computer is the ideal tool for the syntactic operation of characters and the computer can check the syntactic correctness of received input and produced output.

Semantics is the study of relations between text and the objects designated and described by the text, between description and reality. The correctness of semantics cannot be checked, it must be verified against reality—which requires the human mind. Therefore, verification cannot be fully automatized, and there will always remain a gap between any formalized structure and the live nature or the live human mind.

Pragmatics is everything else.

Language reflects and contains the infinite possibilities of the human mind, while formal structures, models which the computer can operate, clearly are of a finite nature, highly combinatorial, but certainly not transcendental.

A system is more than the sum of its parts because of the information flow between the subsystems. A word or a name is more than the sequence of its letters because of its information relationship to other words and to the reality to which it refers. This universe of relations could be symbolized by a cloud hanging over each piece of the text—I am using here a baroque image of which you may think when sitting next time in the theatre—a cloud which appears differently to different people. The computer does not process the cloud, it only transforms character strings. As we can submit certain natural and artificial relations to mathematical expressions, we can as well submit certain aspects and structures of reality to formal models. It has taken three centuries from the Renaissance start of mathematics to reach the universe of present science, technology and economy. We have just started to develop models going beyond what had been formalized since the Renaissance and we will need some time to collect experience in these new areas, to recognize and respect their limitations. Over-optimism is not helpful—as the computer defeat in natural language translation has proved: a defeat, by the way, heavily ignored by the active over-optimists.

I believe that shortcomings, breakdowns and defeats caused by the underestimation of semantic difficulties will remain with us for decades to come. And they will essentially contribute to negative social side-effects of information processing.

But whatever troubles there may occur, the aim will be to include as much as possible in computer-stored models because then all information processing commodities can be applied and enjoyed. Everything outside this computer-oriented universe will not only become costly, it will become strange. Is it not possible that a hostility against uncomputerized items will develop? A hostility against exceptions and against everything that is beyond the algorithmic order?

It is conceivable that information processing will intensify a common philosophy which is already imposed on us by science. We feel today that science can cover, describe, explain and master the full reality; that what is not covered by science is either unfinished work or else an unreal dream, mere imagination, harmful for anyone who cares for it. It is not difficult to imagine a view of the world which follows the same lines, but is further restricted by the computer possibilities: for such a mentality the computer can cover, process and control the full reality; what has not yet been programmed is unfinished work, and what cannot be programmed is either a harmful, unrealistic dream or utter foolishness. This is by no means science fiction, it is simply a logical parallel to the view derived from physics by people who have little notion about the fundamental method and the limitations of the scientific and technological development. They believe in physics rather than in religion—and they prefer to be supported by this pragmatic safety without risks that is inherent in this kind of belief. Why should this commodity not be extended further to the computer—supporting a subset of reality of immensely higher security? You may imagine the details of such a future—a world with almost no human work and effort, a very minor amount of working time per week, vacations during a large part of the year, while information processing systems do everything for us and what they do not do is not worth being done.

Actually, I do not think that this kind of future is the one that will come. I do not say this because I am a confirmed optimist. I say this because there is a further impact of information processing on the general thinking which will more than compensate the safety and fact-oriented tendencies I have just described.

The algorithmic thinking imposed and cultivated by the use of computers forces us to improve planning and forecasting. We will look much more carefully into the future than we have done in the past. And we will get a lot of warnings by doing so—early warnings. The world growth models as initiated and made known by the Club of Rome are a conspicuous example. It is not so important how correct they are at the moment, how well these predictions describe and will describe the future. It is important that the models were possible—possible through automated information processing—that the warning was heard and has found attention.

In this context I came to think of an unexpected impact of automated information processing. In the early days of computers one could have hoped—and many did hope—that by means of the great computer possibilities the science of economy could be turned into an exact science like physics and chemistry. The great men in economics knew better, of course. Oscar Morgenstern, co-author with John von Neumann of the "Theory of Games and Economic Behaviour", has through all his life struggled against this hope, in particular if it was based on this theory of games.

What actually happened over the last decade was not that economics became more of an exact science, but that the science of computer systems became more inexact, in the sense that features of economic uncertainty spread over systems design. It is necessary to plan computer systems longer and longer before they are delivered, and that means that they are planned for an environment that nobody can precisely predict and describe at the time of development. When the system that is planned today will be put in use it may turn out that many assumptions on which it was based are wrong. During the lifetime of this system further ones may substantially change. The non-mechanical nature of our actual environment becomes more and more apparent, and this will undoubtedly have its impact on society.

In particular, as far as algorithmic thinking is concerned,

nobody will ever deny the elegance and usefulness of the concept, but its limitations and its dependency on facts and figures outside the algorithm will be present in the consideration. The necessity to plan and predict the future in spite of many uncertainties will teach us that judgement beyond scientific confirmation is as important in many instances today as it was in the dark ages of the past. The only way to meet the uncertain elements in planning and forecasting is to strive for maximum flexibility of our creations so that they can be easily adapted to unexpected, unpredictable events and trends. If you need scientific proof for this statement then please study the philosophy and reality of correctness proofs, the culmination of algorithmic thinking.

It has already turned out that simple and short correctness proofs are certainly helpful, that most correctness proofs, however, are terribly complicated and lengthy so that they are glamorous candidates for errors, waiting for correctness proofs of correctness proofs. The ideal of algorithmic security appears to be far ahead in the future or even totally impossible.

Enough individuals are not swept away by the excitement about new technological methods—enough specialists maintain a critical view on their own successes. As an example for many I want to mention here Joseph Weizenbaum, the well-known M.I.T. computer specialist who has given in his book "Computer Power and Human Reason—From Judgement to Calculation" a number of highly qualified arguments; indeed, his book is a much better description of many aspects of the impact of information processing on society than I am able to give here.

Joseph Weizenbaum is critical, but not pessimistic. We very much agree on most of the arguments, but in particular on the main principle: the computer is the most important and powerful tool technology has given into the hands of mankind. Any powerful tool includes certain dangers, must have strong positive and negative impact on society. Which of them will be stronger in the long run, the positive or the negative, is in our hands, in the hands of the governments we elect, of the specialists who contribute, and of all the workers in the computer fields who succeed and fail, triumph and suffer, and in the hands of all individuals who benefit and lose, who are helped and hampered by automated information processing.

Nobody can stand aside today and leave the actions to the others—democracy means activity and involvement. Here again the computer can and will have its impact. Have you ever calculated the information flow to the government at your disposal? If there is only one election every 4 years and a decision between four parties, then it is about 64 nanobits per second, really not impressive. I can, however, imagine a computerized system in which the citizen can send much more information up to his government, in which, however, the restriction for any human choice, the power of logical interdependencies and the weight of objective facts is made much better visible than today to the government and to the citizen. Exercising a right involves respect for reality and responsibility for society, virtues not particularly recognized by the present forms of democratic decision-making, but definitely brought to importance by a computerized world in which any kind of public decision-making by vote must depend on proper information, because otherwise the automatic execution of the decision will have catastrophic consequences. The computer-assisted political system has not yet been tried—it is too early in many respects—but I am convinced that it will be tried soon and introduced in the future.

Such an impact of information processing will not mean an era full of luck, happiness and satisfaction. But at least man will be able to decide on the crucial tool of information processing by means of improved information processing. Our fate will remain in our hands, even in a highly automatized technological environment, because information processing will return decision-making into our hands and into the human mind.

A science and technology which is based on information and language, on the highest properties of the human mind can, in the long run, have only one general impact: the return from belief in mechanics to belief in the humanities, in the contribution of the well-educated human mind and in the value of morale and religion.

CHAPTER 13

Development of Communication in the Future Perspective

I. DE SOLA POOL

SINCE the time of Bodin, and certainly since Condorcet's "Esquisse d'un tableau historique des progres de l'esprit humain", men have generally assumed that there is a pattern to human history such that the future will differ from the past. That is a novel thought. Through most of the millennia of human history fundamental change was so slow that men's natural assumption was that their world would not differ greatly from that of their great-grandfathers' or of their great-grandsons'. There were, of course, changes like famines, wars, floods, and epidemics, but they followed no epochal laws or patterns. History could be conceived as "one damn thing after another". Chronicles of these events there might be, but an epic of man with a plot of intrinsic evolution (and not just some magical creation) could only be conceived when individuals could perceive systematic change taking place within the time frame of their own comprehended experience.

The Jewish and Christian myths were among the earliest perceptions of a plan to history. The plot involved creation, sin and the fall of man, a long struggle of penance, and an ultimate messianic resurrection. History, "the unrolled scroll of prophecy", made sense as such a divine plan. Indeed, when in the eighteenth Century the idea of progress was secularized by Condorcet and others, it may have been freed from some of its mythological baggage, but it still had a transcendental character that cried out for explanation.

The Darwinian process of selection was one way to account for a trend of infinite perfectability. Any notion of history more complicated than such unidirectional progress—"a cyclic poem written by time upon the memories of man" as Shelley called it,—was harder to explain except by a divine plan. The nineteenth century notion of scientific laws of history in which "nature" determined the character of society at different stages, misperceived science, and merely substituted the word "nature" for the word "God". Though such sacro-scientific theories may now fail to satisfy us intellectually, they continue, in an era when men recognize the fact of change, to appeal to a profound desire to know where we are going. We are haunted by the question of whither!

The Darwinian explanation of progress makes much sense. It implies that what exists today in the most advanced societies meets some human needs, for it was selected by an evolutionary process. Those human institutions that have evolved over millennia of change may be ones that help the species to survive. But it would be a profound misunderstanding of evolutionary ideas to assume that the process of natural selection necessarily makes life better in all respects for each individual in the species (after all, some species such as bees and ants thrive by their total subordination of individuals). The pleasure principle may play an important role in mediating the process of selection, but that fact does not guarantee that the conditions of life that get selected for survival are the same as those in which we as individuals would find our greatest happiness.

Indeed, it would be very surprising if the features of contemporary society, so alien from the natural environment of our species, were not also to a considerable degree distasteful to the human organism. After all, for almost all of the span of man's existence on earth, people lived in tiny tightly knit primary communities, met perhaps 50 other persons in a lifetime, and talked but little, and that little, mostly in the context of familiar emotional ties where few words were needed. It would be strange, indeed, if organisms that had lived in such a milieu for a couple of hundred thousand years would be entirely at home living in cities of millions, interacting socially not with 50 people, but with two orders of magnitude as many, and spending all but a few minutes a day of their lives in enclosed boxes called buildings.

The society in which we live today is characterized by four quite unnatural major institutions (among others). These are:

the nation state
bureaucracies
cities
and a class structure in which manual labor is performed by an underclass and mental labor by an elite.

These four institutions all grew up together in the past 500 years. Each had its antecedents. Cities existed in the ancient world, long before modern manufacturing and commerce made the city the habitat of the majority of the population. There were city states and empires long before the world was partitioned into (now 130) sovereign nation-states. There were Mandarins and Samurai before there were white collar bureaucrats. There were classes and there was class exploitation, but the members of ruling classes in the past often based their power on personal prowess in combat or on physical possession of land and goods, rather than on moving pieces of paper.

The above four institutions of contemporary society are all well adapted to each other and also well adapted to a communication system using printed mass media. The size of a typical nation state is such that print media can be economically distributed around it in a day, or at most several. A few of the largest nation states are so large that newspapers are based in many separate cities, but more often the national press is the capital's press, published in the major city and circulated daily around the whole country. The unifying language in its canonical form is taught in the schools which are also the source of literacy. The state organization functions through the forms of bureaucracy, built (as Max Weber and others have pointed out) on the written document. Thus those who write, rule and organize the manual labor of the large number of others who make and move things.

Industrial and bureaucratic activities are centered in great cities (and indeed create such cities). In those cities the densely conglomerated population can be well addressed via movies, TV, and the press, produced by an intelligentzia but enjoyed by all.

As Aristotle recognized, the optimal size of the city-state was set by the ability of people to gather together to transact business. One could likewise argue that the optimal size of the nation-state was set by the limits to which printed matter in the vulgate language could be rapidly transported and delivered.

Radio and electronic communication have begun to change that situation. Starting with telegraphy in the nineteenth Century it became possible to transmit the printed word instantaneously over vast distances. However, that was very expensive. It was so expensive that the early newspapers limited their telegraphed text to a short column of cryptic bulletins, leaving the more expository treatments to wait for the mails. But with the coming of radio the cost of addressing millions of people scattered over large areas dropped drastically. The natural perimeter of communication ballooned outward.

In the United States public policy sought to create a broadcasting system modelled on the newspaper system with local broadcasting stations in every town. "Localism" was the basic philosophy of the legislation. It worked, however, only to a limited degree. While no licenses were given for stations with power enough to blanket the country, networking was used by the broadcasters to turn the local stations for many hours a day into mere retransmission points. The vast reach of broadcasting could not be totally denied.

In the 1930s short wave broadcasting made radio a global phenomenon. For the first time frontiers became totally permeable. No customs officer could stop subversive material any more. In Stalin's Russia less than 2% of the population had access to short wave; today the majority do. That fact has been a powerful force for change.

Nonetheless, international broadcasting, international telephony, and other forms of transborder electronic communication are still small in scale compared to the communication within nations. We still live in an era of preponderantly national communication systems. So if we look only at the present world, nationalism seems to be as strong or stronger than ever. We are struck by the extent to which the four institutions that we have mentioned as characterizing the modern world—the nation state, bureaucracies, cities, and the segregation of manual and mental labor—are well adapted to each other, and also well adapted to the main media of communication of the nineteenth and first part of the twentieth Centuries. They are, however, ill adapted to a number of new developments in the world's economy and particularly to its communication system.

We have no crystal ball, but there are a number of trends well under way, of sufficient magnitude that it would indeed be astonishing if they did not continue into the future. Among these we may mention four: the development of a global economy with transactions having externalities that are felt across borders; rapid and efficient transport of persons and things; electronic communication; and intelligent machines. These trends are fundamentally at odds with the institutions that we have mentioned as characterizing the modern world till now.

There is little reason to doubt the power, or the persistence, or the revolutionary character of these new trends. Trade has existed as a global phenomenon for hundreds of years, but the oil crisis of 1973 was the first time in history that the economy of the whole world was instantly and simultaneously impacted by a single set of transactions; before that the crises or transformations that occurred in one place sent ripples or waves spreading sometimes to the ends of the earth. But the oil crisis was not a wave phenomenon; it was a single crisis everywhere.

That fact is explained by the low cost of modern oil shipment. Before modern means of transportation the cost of bulk goods would double in a hundred miles from their origin. Today the shipping costs of wheat or oil almost vanish in the inter-regional differences in their costs of production. The same sort of thing is happening to human travel. Recently, despite the fall of the dollar and America's continental remoteness, one in ten US citizens vacationed abroad. In this era of the Concorde, Jules

Verne's "Around the World in 80 Days" brings a smile of amusement.

The changes in electronic communication are even more revolutionary. Electronic transmissions, as we have already noted, are no respecters of boundaries. Yet up to now economics has kept international telephony and short wave broadcasting as secondary features of the global communications system, but that will not be true for much longer. Three technical changes will result in distance becoming a negligible factor in the cost of communication. The most obvious reason for distance insensitivity of communications costs is the use of satellites. When messages travel to and from a satellite on the geostationary orbit at 36,000 kilometers above the equator, it makes no difference how far apart, within the beam, the point of origin and destination may be.

However, the cost of terrestrial transmission is also becoming more and more insensitive to distance. With broadband digital transmissions, whether over microwaves, through light conductors, or cables, the cost of additional kilometers falls to relatively trivial levels. Finally digital switching and store and forward devices make message processing within a network a complex activity in which distance covered is a minor factor in the cost. In a modern packet network packets travel by random routes, so that on an international network one piece of a message may have travelled through several countries, another piece through several others, and few pieces by the most direct route.

The result of all these technological advances is that global interactions in business, government, science and culture will no longer be merely possible, but also so cheap as to become the usual way of conducting affairs. In that sort of world the nation-state will no longer be able to effectively control communication flows across its borders and therefore no longer be in command of problem solving activities within it.

Increasingly, the important problems in life can only be handled on an international basis. Among the problems that nation states cannot solve individually are pollution, ocean resources, space resources and spectral resources. For example, efficient use of the geostationary orbit has to be transnational, for there are, apparently, large economies of scale. The inappropriateness of national allotment of the geostationary orbit was illustrated by the obsolete political outcome of the 1977 World Administrative Radio Conference at which a plan was adopted giving each nation a spot on the orbit and five frequencies in the 12–14 GHz band for direct broadcasting. The fact of the matter is that there is no conceivable way of avoiding massive congestion on the orbit if each nation puts its own small satellite up there for all the different uses for which fixed satellites will serve.

The only sensible solution is one in which nations share large platforms with a variety of antennas and also switching facilities aboard them. Already a dozen nations are turning to Intelsat to provide domestic satellite communications, while only a few are seriously planning their own national satellites for telecommunications.

The problem of interference between separate facilities will become increasingly acute as high powered beams come to be used to send and receive messages directly from rooftop antennas rather than through each nation's terrestrial communications monopoly. A commercial service of that sort is about to be launched in the United States by Satellite Business Systems. By law, the SBS connections can only be located in the United States, but technically reception is possible anywhere within the beam regardless of boundaries. Satellite interconnections of the future can (and if efficiency is a criterion will) in many instances completely bypass national communications systems as they go from sender to satellite and directly to the receiver.

So the nation state will decreasingly be an effective unit of planning and action in the world that is being created by electronic communication of messages and money, as well as by efficient transport and intelligent machines. Nations will increasingly find that they can no longer effectively control the operations of commerce or the cultural life of their people, as such activities come to be carried on in whatever location is most hospitable, at the will and for the service of people everywhere. Banks (or whatever intermediary financial institutions succeed what we now call banks), will offer their services to anyone who can pick up a telephone anywhere. Journalism, research, and information activities will be conducted globally on distributed computers and distributed data bases. The physical facilities will be located wherever the local authorities are hospitable.

Let me not overstate the case. Local government and community planning did not disappear with the passing of the city state. So governance over large natural regions like nations will not disappear either; but as those matters that sovereign nations can handle by themselves become fewer and fewer, the absolutism of contemporary notions of nationalism and sovereignty will seem increasingly obsolete, and the power of national governments will begin to decline.

Just as trends in communication threaten to erode the power of the nation state, so too they threaten the institution of bureaucracy. Centralized bureaucracy operates by the hoarding and control of information and of its dissemination. In a bureaucracy there are channels that must be followed. A communication system based upon electronic abundance undercuts such a system of restricted channels. The physical capability will exist for anyone anywhere to communicate and share information with anyone else anywhere else.

That process of breaking of communication hierarchies began with the telephone. If one looks at early journalistic treatments of the telephone one finds concern that a woman picking up the phone may find herself talking to a strange man. In some of those articles there is advice to subscribers to have their servants answer the phone to avoid finding oneself talking to social inferiors. An early article in praise of the phone notes how a magnate in Wall Street can talk to the foreman in the bowels of a mine that he owns in Pennsylvania, bypassing the layers between them. When President Kennedy took office he caused shivers to go through the bureaucracy by his habit of phoning assistant secretaries and deputy assistant secretaries that he knew from before, instead of phoning their bosses at the Cabinet level. In the American government today most serious business is carried on by telephone across hierarchical lines, violating all the classic descriptions from Max Weber and others about how bureaucracies work by passing pieces of paper up and down the chain of command. Only in American foreign posts does the old description still hold. Numerous technicians have pointed out to the State Department that modern telecommunications makes it possible to give each foreign service officer abroad a terminal at his desk which would permit him to communicate at will with any terminal back in Washington or elsewhere in a foreign post. The gains in efficiency would be enormous over the present reliance on airmail for most things and on telex or telephone through a bottleneck, for priority messages. But the suggestion is always rejected; the foreign service is clinging hard to the notion that the ambassador is head of the country team and everything must pass through him. It will not be long, however, before this fiction dies and every member of the team finds himself able to talk to anyone back home in the Executive Branch, or for that matter in the Legislative Branch or in the press. Is it any surprise that this prospect evokes alarm?

The statement that anyone will talk to *anyone* else anywhere should not be confused with the statement that anyone will be able to communicate with *everyone* everywhere. Time imposes limits; there are only 24 hours in the day. The telephone removed the physical limit on who could talk to whom, but conventions regarding acquaintanceship, appropriateness, and importance evolved and checked phoning beyond certain limits. In the domestic US bureaucracy, phoning cuts across organizational lines, but officials know the limits as to whom to call. However, the choices of interactions, even if limited, are not constrained by the principles that have been applied in the past. New conventions emerge.

The phone has been with us for 100 years and yet bureaucracy is as rampant as ever. What reason is there to believe that communication technology, for all its flexibility, will reduce bureacracy in the future. The point is that up to now written communication has retained the same canonical form that it had in the nineteenth Century. A document is a considered and fixed representation of an idea. It places a responsibility on its author.

To avoid such crystallization one may use the telephone, but in the end, in a bureaucracy, a document must be produced.

Computers and electronic mail, however, give to documents the same flexibility which a phone conversation has. A memo entered into a word processor can be edited at will by anyone who receives it at his terminal. It can thus have any number of infinitely varied forms none of which need necessarily have precedence over any other. There may never be a printed version which is authoritative. The document may remain a retrievable set of bits in memory, edited and restructured and copied into other memories at will. The implications for copyright, for the organization of science and learning, and for bureaucracy are enormous.

In the ancient world versification was introduced to give some fixity of form to oral sagas that would otherwise have varied infinitely. The notion of sacred scripts was also introduced to preserve texts; any deviation in copying them was deemed a sin. With the coming of printing, secular prose acquired a fixity that it did not have before. In the days of manuscripts, lawbooks, for example, existed in the royal court and perhaps in a handful of copies in major centers. Local magistrates had to interpret the sense of the law from memory. With the coming of printed law books, every court could have the letter of the law at hand. Hundreds or thousands of copies of the canonical text came off the same press in indistinguishable form.

Now we are moving into an era in which text is displayed on a CRT in its fugitive form, and even printout is produced in single copies on demand from the dynamic store of an electronic memory.

If we believe that electronic communication is promoting the atrophy of nation-states and the atrophy of bureaucracies, can we believe that despite that there will be a world of great cities. One might imagine a new Hanseatic League tied together by verbal and material commerce, but that seems unlikely. Co-location of people (i.e. cities) loses much of its importance if remote communication is easy. We can anticipate that interaction with people at a distance and interaction with those close at hand will become of much more nearly equal efficiency. Work done by machines, rather than by human hands, will be easily controlled from a distance.

The pattern of densely populated cities with sparse agricultural areas between them is already giving way to complex megalopolitan regions. Cities, before the invention of telephones, streetcars and automobiles, were walking cities. One had to be able to go from one end to the other on foot. To reduce the duration of long walks different activities concentrated in small neighbourhoods. A city was a checkerboard of single-activity neighbourhoods, still signalled, in old cities, by names such as Market Street or Ironmonger's Lane.

With the coming of public transport, cities spread, and with the coming of the telephone, entrepreneurs moved their businesses to cheaper locations, and also nearer to their customers, without losing contact with their suppliers, bankers, and labor pool. In the first part of the twentieth Century the downtown became a center of offices and department stores, while production moved to the outskirts. To meet the demand for office blocks, skyscrapers were put in the heart of the downtown. Without the telephone, as well as the elevator and steel-frame, these tall buildings would not have been possible. In 1902 someone figured out that if businesses had had to continue using errand boys instead of phones in skyscrapers, the traffic would have required an elevator well of such size as to make such tall buildings uneconomic.

So a dual process was taking place of urban concentration along with spread of both industry and residence to suburban periphery. Instead of cities we now have in regions like Bos-Wash or the Ruhr a network of population concentrations extending over hundreds of miles. There is every reason to expect this process to continue with an ever more complex division of labor among centers of population density.

Urban concentrations will not end any more than will nations or bureaucracies, but they will be transformed perhaps beyond recognition. Population projections for the year 2000 tend to show that virtually all the growth of the worlds population will be in urbanized areas. The rural population of the world may perhaps remain fairly constant, and so there are likely to be then, as there are now, vast areas of forest, desert, fields, and wilderness. But the urbanized areas may be less distinguishable into finite cities, and very likely there will be vast areas of exurbia in which people will seek an urban way of life in a rural atmosphere. Abundant communication may make that possible. Cities and exurbia each have their attractions, but the balance between them is profoundly changed by telecommunications and the changed structure of work.

The structure of work has already changed in the most advanced societies to the point where the majority of the working population are workers with brain rather than with hand. The documentation of that remarkable trend is found in the work of Mark Porat on the United States. His study is now being replicated under OECD auspices in various other countries. In 1900, according to Porat's figures, 13% of the U.S. labor force were in information occupations as against 27% in industry. By 1950 31% of the labor force were in the information sector, by 1960 42%, and by 1970 46%.

That is a revolution! Stop and think for a moment about those figures. In a quarter of a century, the proportion of the labor force moving paper or other information media instead of things has gone from one fourth to one half of the total labor force.

That extraordinary rate of growth in the information sector will not continue. The large number of white collar workers makes it now economically attractive to develop labor-saving devices for intellectual activity, in the same way that in the last century when production required large masses of manual workers, it paid to invent labor-saving devices. Just as large numbers of persons, including many who are no longer manual workers, use machines to do things which used to require manual labor (e.g. a washing machine or a car), so in the future we may expect large numbers of people to use thinking machines, though not necessarily to be themselves employed as white collar workers. What occupations can we expect to grow in a society in which the agricultural sector in the U.S. has declined to 4% of the labor force (a figure incidentally for a very productive agricultural country), and in which the industrial working class may be expected to decline to a similarly tiny number as automata take over assembly lines, and in which clerical work too begins to be automated by computers and so comes to require a sharply reduced part of the labor force. The jobs that will remain to be done by human beings are those that require a specifically human kind of interaction. These are service jobs such as those performed by a waiter, or nurse, or counsellor, or teacher. Such person to person services are likely to absorb an increasing portion of the labor force. We now know how to build machines to automate the more routine intellectual as well as manual functions, but we do not know how to automate the experience of a human relationship.

If what we have anticipated as a consequence of electronic communication is even for the most part correct, we are at the edge of a vast transformation of society. We seem to be moving towards a society in which nation states will lose much of their power, in which bureaucracy gives way in part to more equalitarian and distributed modes of decision-making, in which cities are replaced by megalopolitan regions, and in which only a small minority of the population remains engaged in the classic forms of productive toil that have been the lot of most of mankind until now.

These prospects of drastic change have very little in common with most political platforms that advocate drastic change. Most movements that are self-described as radical are highly urbanistic, or nationalistic, or oriented to obsolete class structures, or to central bureaucratic planning. The changes that we can see on the horizon are much more drastic than that. They are changes that arise out of the very nature of electronic communications technologies. They reflect the ease with which electronic communication can operate over global distances, and the abundance of bandwidth that can now be made available to all, without producing any exhaustion of the earth's resources. They reflect the ability of digital communications to operate under the control of thinking machines. They reflect the ability of intelligent machines to handle degrees of complexity far beyond anything conceivable with non-electronic devices, and at the speed of light. These technological facts have profound social and political consequences. People who think about social change in traditional political terms cannot begin to imagine the changes that lie ahead. Conventional reformers cast their programs in terms of national policies, or in terms of laws and central planning. But in the end, what will shape the future is a creative potential that inheres in the new technologies of electronic communication and machine intelligence.

CHAPTER 14

Do Computers Lead The Way to a New Social System?

F. MARGULIES

WE are living in a period of ever accelerating quantitative and qualitative technical development. One or two decades ago, technical and scientific progress was greeted with great expectations and enthusiasm. Today, this attitude is changing to one of anxiety, fear and even open hostility. Not only the use of atomic energy but the unplanned and uncontrolled development of technology in general is being looked upon as the cause of all evil in our world.

Professor Thomas B. Sheridan of the Massachusetts Institute of Technology, in a paper presented to the Joint Automatic Control Conference at Purdue University in 1976, painted a rather gloomy picture:

> "The history of man has seen a gradual diminishing of man's self image: Copernicus taught us that man is not the center of the Universe; Darwin taught us that man is not different from animals; Freud reduced man's stature still further by pointing to his irrationalities and inconsistencies. Will the computer-robot and its alienating influences be the ultimate insult to his dignity?"

As early as 1556 Georgius Agricola, in his comprehensive work "De re metallica", felt obliged to refute the argument put forward against mining and metals: i.e. that God had pushed the metals into the depth of the earth so they could not be molested by man; and that mining, therefore, was an offence against the will of God that would bring woe to man. Agricola refutes this argument: when metals are used for destructive purposes, it is not that they are to blame but our own vices, our rage, our discord, our greed. He wrote: "Here the question arises whether what we dig out of the earth is to be classified as good or bad. Good men use it well and it is useful to them, but bad men use it badly and to them it is useless. But if someone uses it for an evil purpose, it cannot be justly called evil in itself. After all, is there any good thing that we cannot equally use for good or evil?"

Life is permanent interaction with nature, the dialetical struggle with an environment, which in general creates life and makes it possible, while individual existence is permanently put in jeopardy by these very same phenomena. Plants and animals, directly confronted with this environment, cannot but react passively to it. Their weapons in this battle are of a biological nature, and the nature of the battle only changes in consequence of biological changes, i.e. over periods measured in thousands or millions of years. Their chance of survival consists not in changing nature but in adapting themselves to nature.

Man is the only animal to have invented tools to deal with nature and, with the power of his intellect, to have transformed the struggle with nature into the process of work. Man confronts nature not with weapons developed by nature but with artificial means created by himself.

These weapons are tools, machines, the ability of man to put them to use, the social relations and the structure of society, or, generally speaking, the stage of development of productive power and production relations. The stage of development of these weapons determines the relationship between man and nature. These forms of struggle are being changed not by biological but by technical and social developments which take place not over a period of millions of years but within decades and, these days, very often within years. Thus, ever more frequent changes are forced upon man and upon society: the development of productive forces is the real motor of the development of society—

this contention of Karl Marx today is as true as ever.

In his relationship to nature man is characterised by his activity and by his creativity which is transforming nature. By exploring, understanding and applying for his own benefit the laws of nature, he has made "the earth his servant".

Therefore it seems all the more peculiar to encounter, time and again, waves of emotion against science and technology in newspapers, books, speeches, in conversations in bars and pubs and even in scientific analyses. Are we really sated with technology and progress, or are we just thoughtlessly following a passing fad? After all, nobody really wants to give up those achievements of science and technology which have been attained with so much difficulty; nobody wants to return to life as it was 100 or 200 years ago, without sanitary installations and sewers; nobody wants to go back to steam locomotives, to the dwellings and to the working conditions of those times; nobody would prefer the sanitary conditions and the life expectancy of around 1800 to the conditions prevailing today. Nobody would be willing to build houses and roads, to produce iron, steel or energy with the methods of the nineteenth century.

Certainly, technology must not become an end in itself. Abuse and misdirected development must and may be prevented. However, we are far from a point of saturation, or even super-saturation with technology. Nobody can deny that there is enormous need in the world as a whole—need which in many cases is increasing. Even developed and rich countries lag considerably behind in various fields: I am thinking of education and health, of housing comfort, of traffic, of the elimination of poverty and of complying with necessities of private consumption (and not with artificially created demand), of the improvement in working conditions, of easing heavy work and eliminating monotonous work.

No doubt, it will take decades to eliminate these deficiencies and to fulfill these desires. And then there will arise new desires and new needs requiring fulfilment. They will not necessarily be of an economic nature, but they will be based on economic security. A standstill of wishes and needs would mean a standstill of human development, something which hardly anybody would want.

But if work—the employment of intellectual capacities and the conscious production and use of tools—constitutes the dividing line between animal and man, this means that work is of paramount importance also for the future of mankind. Work is not only a concept of physics, known by heart to every high-school student. It is much more than that: work is a philosophical, a sociological, a psychological or, to put it in a better and more comprehensive manner, a humane activity. There is no work, however automated it may be, which does not implicate man at some stage, there is no product which can be made completely without man. All the wealth of this world is the product of human work, and, vice versa, mankind could not exist without work.

By his unceasing endeavour to make work better and more effective, man has developed new means of production as well as new ways of human coexistence and cooperation complementary to these new tools.

Prehistoric society was adequate for the primitive tribal community of hunters and gatherers: Everybody was equal (and nobody more equal than another!) and everybody lived from hand to mouth which meant that everything produced was immediately consumed.

The hunting weapons—and therefore the "productivity" of hunting—improved: fire and salt were discovered and with them "technologies" of food conservation developed. Thus an economy of storage developed, and with it a social class emerged which had to assure the formation of stocks ("foregoing of consumption") and their administration.

The development of agriculture and animal husbandry was matched by feudal society which brought the division of society into land owners, serfs and artisans.

Capitalism finally was based on the industrial revolution in the eighteenth and nineteenth centuries—on steam power, on engines for work, on the organisation of factories, on mass production and division of labour, on Taylor's "Scientific Management", on Ford's conveyor belt: in short, on a gigantic development of productive forces and a simultaneous lag in the individual development of man. By splitting work into smallest components, by constant exact repetition of these components and by thus extorting a maximum of output while reducing work content to a minimum, man was degraded to a machine, to an automaton. In the process of such work he could not make any decisions; every move was regulated by instructions; the machine, his master, did not belong to him; the product, in whose production he participated, belonged to others. This way the worker lost his personality, the personal relationship to his work.

Beginning with the end of the Second World War the industrial nations entered a new stage of technological development—the stage of scientific-technological revolution. It is a stage characterised by a comprehensive development of all factors of production: automated methods, new organisational techniques, new materials, new sources of energy. It is furthermore characterised by the transformation of science into a direct productive force, by the application of mathematical methods to many branches of science which formerly were not amenable to them, by the application of systems theory and of modelling and simulation in experimental research, by the rendering of production "scientific", by an ever closer liaison between theory and practice.

This stage is—last but not least—characterised by the growing importance of man, in spite of automation and computerisation. Division of labour, man as an appendage of a machine, these principles are no longer in accord with the new technology. The more complex a system is, the more it is dependent on intellect, creativity and the human ability to make decisions.

The scientific-technological revolution of our times is also characterised by a speed of development previously unknown in the history of mankind. In 1946 the computer was a hardly manageable giant, not much more than a huge toy for eccentric scientists. Ten years later the triumphant advance of electronic data processing (EDP) in all branches of the economy started, and in 1960 more than 11,000 computers had been installed in the USA alone. Today, there are well over 100,000 in use. Simultaneously, the criteria of performance could be increased 1,000-to 5,000-fold. If, after the pioneer work of the Wright brothers, aviation had developed at a similar speed, the landing on the moon would have been possible in about 1920.

However, the more technological development accelerates, the more urgent—but also the more difficult—becomes the adaptation of man and of society to it. The steam engine was invented about 80 years before it was generally used in industry. The period of time between the invention and the general use of the electric motor was about 30 years. The time of a generation thus could be used for adapting man and society to these machines. The computer gave us only a few years time. Thus, technologically we are living in the twenty-first century, but are still far back in the nineteenth century so far as our social institutions are concerned. For instance, in my country, Austria, the instruments and the institutions of parliamentary democracy were shaped at a time when the motor car and the telephone were just being invented and when the ability to read and write was considered part of higher education. Today, with our ballot system we still authorise men and women whom we consider likeable and trustworthy to govern us for four years; should we want to influence their decisions during this time, we would have to start a very complicated and slow procedure. A modern computer enables us to make a few hundred thousand decisions a second. Modern democracy, on the other hand, permits us only one decision in 126 million seconds (i.e. in 4 years).

The computer is to be regarded as the symbol and pacemaker of the scientific–technological revolution. What is the significance of this revolution for the development of our society?

In reality, every man-machine-system, especially a man-computer-system, links two extremely contradictory structures.

The computer is an artificially-created, formalised system with stable features. Once having been designed and built, it can fail, but it can never change. Man, on the other hand, is a natural, non-formalised system. He is never stable, he changes as long as he lives—even if he does not fail.

A computer can be completely described with drawings, figures and wiring diagrams. Any expert in possession of such a description can reproduce this computer as often as he wants to, and it will always possess the same, previously-determined qualities and parameters. Should you try to describe a human being in the same way, you will soon find out that the more copious and detailed the description becomes, the less it will provide a picture of the personality. Reproduction is completely impossible.

As far as performance is concerned, the computer more closely approaches human capacities than any other machine, but more than any other machine, in this performance the computer depends on man. The computer does not only need man, it also develops him, and this developed man in turn makes his imprint on society.

If a computer is grafted upon an organisation designed according to Taylor's "Scientific Management", in most instances the consequences will be very negative. The reduction of human work to a perpetual reiteration of the same movements; the non-utilisation of man's intellectual abilities—however they may differ; the early termination of man's process of enlarging his knowledge and his abilities, since this kind of work does not give him a chance to learn anything new; all this is not only an offence to human dignity but also an offence to the system.

Norbert Wiener, the father of Cybernetics, wrote in his book "The Human Use of Human Beings" more than 20 years ago:

> "In my opinion, any utilisation of a human being is degradation and waste if it asks for less and expects from him less than his full capacities. It is a degradation to chain man to a thwart and use him as a source of energy; but it is almost as great a degradation to set him in a factory to a perpetually reiterating task which claims less than a millionth of the capabilities of his brain."

The computer—and with it the new technology—brings about new standards. The problems of a computer system cannot be mastered by man who is just an appendage of the machine, or by thoughtless button-pushers. They can only be coped with by men who think, make decisions, act consciously and are able to intervene if routine and programs fail. Especially the new developments of computer architecture, characterised by key-words like "distributed processing", "distributed intelligence" etc. point in the direction of a non-Taylorised work organisation. Intelligent terminals simply need more intelligent users who can not only handle the terminal but who also have the capacity and the power to take those decisions which the "computer at the place of work" makes possible. This way, higher and more comprehensive education becomes a "must" in our time. The computer demands it, as the steam engine demanded general basic education or as the development of trades brought the strict guild rules and the training of journeymen.

"Knowledge is power" and "education makes free". Once the powerless and unfree wrote these slogans on their banners, so scaring the already powerful and free. They were afraid of education, scared of those who were educating themselves.

From their point of view their fear was not quite unjustified. The journeymen tramping the roads in far-off countries brought home new ideas and rebellious thoughts. They forced the nobility and the masters of the guilds to make concessions which finally led to the introduction of constitutional monarchy. The children of the workers, having learned to read and write in primary school, showed their dissatisfaction: they fought for more rights and they finally gained the general franchise. And now participation in decision-making at the place of work is a topic of the day.

To be sure, automation in factories has far-reaching effects on economic policy. Therefore participation and control are necessary consequences. The citizen has become increasingly aware that the catastrophic cycle of senseless quantitative increase in consumption and a just as senseless increase in production and productivity must finally be controlled. Quality instead of quantity is the slogan which is being increasingly accepted—generally, and not only in energy policy.

Cybernetic thinking takes place in extremely abstract terms:

mathematical formulae (algorithms) as descriptions of complicated real processes, the computer as a ''black box'', interesting only for its input and its output, etc. In the fields of data-processing and of control technology this method of thinking has become a matter of course among experts; in daily life, however, everything is the other way round. ''Eternal truth'' is being preached while continuous changes are taking place, the ''best way'' is reaffirmed while only a choice between continuously emerging new alternatives can be of help; facts are being manipulated to serve ones obstinacy while only feedback could lead to an approximation to truth. But if cybernetic thinking and systems thinking could be introduced into our economy or our political life, the consequences for individuals as well as for the development of society would be of extreme significance.

Let us return to the question posed in the heading of this paper: Do computers lead the way to a new social system?

I hope I have made my point: I view the development of productive forces, the development of technology and of man, as the real motor of social development. I view the computer as one of the most important, most revolutionary tools ever created by the human race. I view the previously-described effects of the computer as important beginnings of social changes.

But all this will not come about automatically; it is not preordained. On the contrary, the computer gives man enormous new chances. It enables him to determine the course of his life himself, to get information first-hand, to make sensible decisions. But the computer can only furnish the basis for these decisions, it can never decide itself.

''God created the nuts—but we ourselves have to crack them'', the poet says. This means, applied to the subject of this paper: *The computer does not lead us into a new social system. Only we ourselves can accomplish this.*

CHAPTER 15

Communication and Understanding: A Computational Approach to Language Behaviour

R. NARASIMHAN

Introduction

Animal and human communication

One of the principal themes of the DISCOVERIES Project is "Communication" from the human viewpoint. I have chosen to discuss language behaviour and some central issues that arise in trying to model natural language behaviour in psychologically relevant ways. Although not all human communication is language-based, language undoubtedly is what gives human communication its enormous power, versatility and effectiveness. Medawar* has persuasively argued that "what is characteristic of human beings considered as animals is not the making and use of tools but the communication from one person to another, especially in the next generation, the knowledge and know-how to do so". It is by virtue of this faculty that human beings come to enjoy a kind of cultural evolution which, Medawar points out, is, in contrast to Darwinian "endosomatic" evolution, "Lamarckian in style, that is, it embodies a learning process; for what is learned is passed on and becomes part of the evolutionary

* P. B. MEDAWAR: Ethology and Human Behaviour. In: *Growing Points in Ethology* (P. P. G. BATESON and R. A. HINDE, Eds) pp. 497–506, Cambridge University Press, U.K., 1976.

heritage''. Also cultural heredity is mediated through non-genetic channels, and it is also reversible. And Medawar notes, ''this characteristically human system of heredity calls for and depends upon the existence of language and other forms of conceptual communication''.

Although communicative competence is something that human beings share with other animals, the implication of Medawar's argument is that this communicative competence seems to have resulted in cultural evolution only among human beings. The reason for this would seem to be the crucial role played by language as a medium of communication among human beings. Language behaviour is a modality that seems to be functionally available only to human beings. Although speech is the most important constituent of language behaviour, clearly, it is not indispensable. The deaf and dumb are able to use gesture (or sign) language with almost equal efficiency. In view of this, the first question that calls for an answer is: why is it that no other animal has developed language behaviour for use? Perhaps, to answer this we have to ask another related question: what minimal organization of behavioural mechanisms is necessary to acquire and use language behaviour? It seems plausible that we would be in a better position to tackle this second question if we could first determine, what types of behaviour, otherwise unavailable, are made available through language usage. Behaviourally, in what sense do human beings differ from all other animals because of the availability of language behaviour? What is it that human beings can accomplish that other animals cannot, and that is predicated on the availability of language behaviour?

Concerning this last point it would seem that Pavlov's* insight, that language behaviour is a *second signalling system*, really provides the basis for characterizing its role. Without the language modality one has merely the sensory-motor behaviour complex that directly relates to the external world and can only deal with the ''now'' and ''what is out there''. But with language behaviour one is able to separate out the sensory and motor

* I. P. Pavlov, *Experimental Psychology and Other Essays*, Peter Owen, London, 1958.

complexes and mediate them via the language behaviour schemata. That is, one is able to *refer to* the sensorily apprehended aspects of the world and one's own action schemata independently, and *talk about* these using the language modality. In this way language behaviour becomes a second signal, a signal of signals. However, as Pavlov* has emphasized, "it cannot be doubted that the fundamental laws governing the activity of the first signalling system must also govern that of the second, because it too is the activity of the same nervous tissue".

Animal communication seems to lack this second signalling aspect except in extremely rudimentary forms. Vocalization is made use of to indicate need states (say, hunger), affective states (say, anger), and for purposes such as warning of predators, ensuring mother-child bond, maintaining social cohesion, announcing food, and so forth. Thus, vocalization serves to manipulate the environment as well as to describe it, in a certain fashion. However, two important points deserve to be emphasized. Vocalization always relates to situational aspects (external and/or internal) that are present here and now. We do not have any evidence that an animal uses (or, can use) vocalization to warn another animal (say, its young one) of potential threats: e.g., "when you go near the water-hole, watch out for tigers".

Secondly, the descriptions of situational aspects conveyed by vocalizations lack structure. It has been claimed that vervet monkeys have three different kinds of alarm calls: one for hawks, one for snakes, and the third for other terrestrial predators. It has also been suggested that the variations in the way a given songbird sings may be correlated with the various activities the bird engages in. This would enable the listening mate to monitor some of the activities of the hidden singer†. But such subtler aspects of descriptions would seem to be exceptions. In a true language, a description would enable one to discriminate between "a lion", "a big, fierce-looking lion", "the big, fierce-looking lion that we

*See previous reference.

†W. J. Smith: Message-Meaning Analysis. In: *Animal Communication* (T. A. Sebeck, Ed.) pp. 44–60, Indiana University Press, Bloomington, 1968.

saw yesterday'', and so on. There is no evidence that animal communication is based on such structured descriptions.

These considerations lead us to repeat our earlier question: why is it that animals lack language? Lieberman* concludes from a detailed spectrographic analysis of a variety of primate vocalizations that, (1) these animals do not move their tongues during a cry, and (2) their laryngeal outputs tend to be aperiodic. According to him, their vocal apparata are not adapted to producing human speech because they lack laryngeal control and a pharyngeal region capable of varying its cross-sectional area. The several unsuccessful efforts that have been made to teach speech to chimpanzees would seem to confirm Lieberman's views.

However, as we remarked earlier, speech is not indispensable to language behaviour. Geschwind† conjectures that the ability to develop language in man is very likely the outcome of two features: (1) that the major part of the human brain consists of association areas directly linked to the primary sensory and motor areas, and (2) that these association areas are themselves richly interconnected through a centralized association area of association areas. In other animals, including the higher primates, even the primary association areas are not present (or, are only poorly developed), and the sensory and motor areas are only linked to the limbic area (concerned with behaviour immediately relating to survival). Geschwind's claim that higher primates cannot accomplish intermodal transfer (not involving the limbic system) has been shown to be not valid‡. The Gardner's success in teaching sign language to a chimpanzee would also seem to establish convincingly that non-limbic intermodal transfer is possible at least among the higher primates. Geschwind's main conjecture could still be true so far as spontaneous language acquisition is concerned. As we shall see

* P. Lieberman: Primate Vocalisations and Human Linguistic Ability, *J. Acoust. Soc. Am.* **44**, 1185–87, 1968.

† N. Geschwind: The Development of the Brain and the Evolution of Language. *Georgetown University Monograph Series*, vol. 17, pp. 155–69, 1964.

‡ R. K. Davenport and C. M. Rogers, Intermodal Equivalence of Stimuli in Apes, *Science*, **168**, 279–80, 1970.

later, the model that we develop is also based in an essential way on the availability of associative connections between the various modalities.

Even if animals do not acquire language on their own, the question remains whether they could be taught language through systematic training. The most widely known attempt to answer this question is, of course, that of A and B Gardner's, of the University of Nevada, to teach the American Sign Language to a young female chimpanzee named Washoe*. At the end of 3 years of training, Washoe could reliably use close to ninety different vocabulary items of sign language. These included signs denoting objects, commands or requests, attributes and proper names. She could construct meaningful sequences of such signs (e.g., please tickle more, come Roger tickle, out open please hurry, you me Greg go, etc.). After having been taught the sign for 'dog', she used it appropriately to refer to pictures of dogs and, once, even to the barking of a dog not in view.

Although the communicative competence displayed by Washoe using the sign language medium is certainly impressive, there is one aspect in which Washoe and a human child differ significantly in their use of language. Washoe's use of language still relates only to the "here" and "now". Practically all her utterances deal with her own desires and needs, with what she sees and hears. There is little indication that she ever uses language to describe impersonally what other people around her do. Also, she does not use language as an instrument to explore. The Gardners do not say that she ever asked her human companions, "what is this?" or "what is that?"† Children at a comparable developmental stage undoubtedly begin to use language to explore, not only the world around them, but various imagined possibilities and

* R. A. Gardner and B. T. Gardner: Two-way Communication in an Infant Chimpanzee. In: *Behaviour of Non-human Primates* (A. M. Schrier and F. Stollnitz, Eds) vol. 4, pp. 117–184, 1971.

†More recent information concerning Washoe's language behaviour development may be found in R. Fouts *et al.* Studies of Linguistic Behaviour of Apes and Children. In: *Understanding Language through Sign Language Research* (P. Siple, Ed.) Academic Press, New York, 1978.

potential situational aspects; for example, "what would happen if I do this and that?", "why not this and that?", etc. There is no indication that Washoe ever attempted to use language in this way to create a world for herself, to fantasize or mythologize. Clearly, these are some of the very early uses of language that children make, in addition to using language in monologuing and role-playing.

After 5 years of training, I believe the experiments with Washoe had to be discontinued because she had grown too big and potentially too dangerous to work with. She was hence removed to a primate colony. Unfortunately, I have come across no published information of Washoe's subsequent language behaviour history among her chimpanzee companions. It would be of great interest to know whether her training in language had given her the motivation to act as an agent of cultural evolution in her primate colony.

Human use of language

The distinguishing feature of human language behaviour is that through its use one is able to deal with a world not necessarily immediately present to the senses. Thus, one is able to perform *gedanken* experiments instead of being limited to performing only *actual* experiments. Once this capacity is reached, one is able to talk about the talking-about aspect itself. In other words, language modality for the first time becomes capable of self-referencing. This gives rise to the object language—metalanguage distinction. This also gives rise to the well-known semantic paradoxes of logic and the foundations of mathematics.

Primitively, however, language behaviour is used only in the instrumental mode. That is, it is used to describe and manipulate the external, non-linguistically given, world of objects and events. The central computational problem involved in modelling language behaviour is concerned with the methodology of mediation. How is one enabled to make use of language behaviour to talk about—to describe, command and query—the world normally dealt with through the sensory-motor modalities?

How does a child manage to acquire this competence to use language behaviour in these ways?

The more recondite use of language behaviour is, of course, associated with cognition, with the highly-complex, symbol-manipulation activities. Complex problem-solving, the construction and use of theories, and similar overtly syntagmatic activities would presumably be impossible without the use of a language of some kind. However, one should approach the complexities of language behaviour with some caution. It is very easy to be misled into the belief that the seemingly unlimited complexities of language behaviour necessarily imply a comparable complexity in the underlying behavioural mechanisms. It is worthwhile to note that the recondite syntagmatic aspects of human behaviour are untypical. They are based on prolonged and systematic learning and training. It is an exceptionally small proportion of human beings that come to exhibit these aspects of language behaviour with facility. The behavioural mechanisms underlying language behaviour can only be articulated by studying what is typical of language behaviour among all human beings. This is, as we said earlier, the use of language behaviour in the instrumental mode which is related to its ability to mediate sensory-motor behaviour. So we must answer the question, what are the primitives of language behaviour that enable its being put to this mode of usage? Given these primitives, we could then proceed to investigate how systematic training procedures could enable them to be used to realize the symbolically more complex forms of language behaviour?

In the next section we shall study the instrumental aspects of language behaviour in some detail. We shall try to analyze what does interaction with one's environment in the language modality consist of. That is, we shall be concerned with an analysis of natural language behaviour in order to be able to answer questions of the following kind: in the instrumental mode of behaviour, how is natural language made use of? What is there in natural language which allows it to be used in these kinds of ways? Since, both phylogenetically and ontogenetically (in the case of humans), an organism's ability to interact with its environment is acquired, to begin with, without the mediation of the language modality, we should develop a framework in which we can first talk about an organism's interaction with its environment in a neutral sense independently of specific modalities. This would then allow us to discuss how, when language behaviour is acquired, it is enabled to assist, mediate, and ultimately replace such overt and immediate interactions.

These arguments would lead us to the view that there are definable aspects to a natural language that enable the members of a language community to use it the way they do. We shall then argue that acquiring a language consists essentially in acquiring these aspects of the language and relating them to their usage. The validity of this thesis can be tested in two ways. Firstly, we could try to analyze some particular natural language with respect to its definable aspects as formulated in our thesis. This would enable us to demonstrate, both in terms of structure and function, how natural language behaviour is enabled to play its second-signalling role. Elsewhere I have attempted a partial analysis of English with this end in view*. Secondly, we could try to validate our thesis by considering its implications to child language acquisition and studying to what extent these implications are verified by known empirical facts relating to the developmental stages children go through in acquiring language. In the report cited earlier I have attempted this analysis also. However, I shall not repeat either of these here since they would take us too far outside the scope of this paper.

I shall confine myself here to discussing the relevance of our analysis of language behaviour to the structuring of computational models to simulate aspects of this behaviour. Our primary concern will be with the kind of constraints a viable simulation model should satisfy. I shall argue that currently prevailing grammar-based models of language-understanding cannot cope with the developmental aspects of language acquisition. In contrast to grammar-driven syntagmatic processes

* R. Narasimhan: *Modelling Language Behaviour*, Technical Report, Computer Group, T.I.F.R., Bombay, 1973.

that form the foundation of computer programming, we shall be led to the view that example-driven (i.e. paradigmatic) processes such as imitating, analogizing and rehearsing should form the basis of natural language acquisition and usage. I shall illustrate some of these ideas with the first version of a Paradigmatic Language Acquisition System (PLAS) that has been implemented.

The Mediating Role of Language Behaviour

Organism as a behavioural system

In trying to discuss what role, primitively, language plays in behaviour and how it is able to play this role, we must first of all note two fundamental points. A child acquires its language through intimate behavioural interaction with a language community. In other words, it acquires a language by *functioning in the language* as part of its normal process of growth. It should, hence, be reasonable to suppose that the developmental aspects of language behaviour should reflect both the developmental aspects in the growth of the child, and the developmental aspects of the language learning process. Secondly, a child, that is capable of acquiring and using language in this manner, when viewed as a behavioural system has certain well-defined and characteristic features. It has inputs coming in from the external environment through its sensory interfaces: eyes, ears, nose, skin, etc. It has certain kinds of motor actions available to manipulate the external world. There are well-defined internal state variables that define the internal state of the system; for example, the affective states, need states, and so on. These characteristics define various aspects of the world available to the child considered as a behavioural system. These must be mirrored in some sense in language behaviour if it is to be possible for the child to function in the language modality to accomplish various things in the world.

We shall try to develop now a somewhat formal view of a child as a behavioural system taking explicitly into account the two basic points outlined above. This would enable us to relate language behaviour to the universals of behaviour, in general, and discuss how language is able to play its mediating (i.e., second-signal) role.

We first note that an organism is always embedded in some (stimulus) situation and, hence, its behaviour must be characterized relative to this situation. In doing so, however, it is the *total* situation that must be taken into account. This total situation can be broadly divided, in an obvious way, into that aspect of the situation that is external to the organism (the external situation, or the so-called *external environment*), and the aspect that consists of the organism itself (the so-called *internal environment*). Let us now make the following postulate.

> P1: Behaviourally, the organism consists of a set of external and internal interfaces. Various aspects of the total situation are defined (or determined, or realized) at these interfaces. From the point of view of the behaving organism, its behaviour is related to the total situation through the aspects of the total situation as available at these external and internal interfaces.

What is being asserted by P1 is, of course, that so far as the organism is concerned, its "awareness" of the total situation is *completely* characterized by the aspects of the situation available to it at its various interfaces. This postulate would be pointless if all that exists in the entire world is a single organism. For, then, it would be unnecessary to distinguish between a "total situation" and "what the organism is aware of". Implicitly, then, we are assuming the existence of a plurality of organisms in the world and that some of these could *observe* the behaviour of others. A behaviour theory could be meaningfully developed only in this larger context.

Our next postulate concerns the *representations* of these situational aspects as available at the interfaces.

> P2: The aspects of the total situation at the interfaces are given in terms of schemata and/or state variables.

"Schemata" is used as a generic term to refer to abstract structures. A specific schema delineates a particular structure. A schema could represent a spatial structure, a temporal structure, or a spatio-temporal structure. We shall consider some examples presently. The distinction between schemata and state variables is pragmatically an important one. Schemata imply structure; state variables refer to unstructured parameters that characterize certain aspects of the situation and, hence, of the behaviour which is a function of these.

To illustrate the ideas implicit in P1 and P2, we might consider some examples of interfaces and the related situational aspects determined as schemata and as state variables.

(1) Sensory Interfaces
a. visual : shape schemata.
b. auditory : schemata for speech syllables, say.
(2) Motor Interfaces
a. hand movement : schema for writing the letter 'A', say.
b. leg movement : schema for walking.
c. face muscles : schema for 'looking angry', say.
(3) State Variable Interfaces
a. colour sensors.
b. taste sensors.
c. parameters of internal regulatory processes; e.g., CO_2 concentration in blood, pulse rate, etc.
d. voice pitch, loudness, etc.

Some of the external interfaces relate to language behaviour and, hence by P2, there must be schemata and state variables asscociated with these. Let us refer to these schemata as language schemata. Our principal aim in the rest of this section is to discuss the nature of language schemata and consider how they are related to the non-linguistic schemata and state variables which define specific situational aspects.

Normally, the primary interface associated with language behaviour is the motor interface relating to the speech mechanisms. However, there are also several secondary outputs (like facial expressions, gesture, etc.) that play a more or less determining role in language behaviour. Thus, apart from the language expressions proper, these secondary outputs, as well as other prosodic aspects of speech such as intonation, stress, pauses, and so on, should be considered as forming part of the complete language schemata. In our subsequent discussions, however, we shall restrict our considerations to the linguistic expression part of language schemata for reasons of simplicity.

We have already emphasized that speech is not indispensable to language behaviour. "Language" behaviour in such non-standard cases must be appropriately interpreted. It would presumably be found in these cases that what we have characterized above as secondary outputs, as also schemata associated with some of the other modalities, come to be used as substitutes for speech and hearing.

P1 and P2 have been concerned with the manner in which the situational aspects are made available to the organism and how the organism is enabled to act. We need a last postulate to characterize from a behavioural point of view these acts.

P3: The universals of behaviour are (a) describing situations, (b) manipulating situations, and (c) exploring situations.

Describing, manipulating and *exploring* are meta-behavioural categories. That is, they apply equally to behaviour in any modality. One can explore with the eye as well as with the ear. Aspects of some situations can be described with facial expressions as well as with hand gestures. But, quite obviously, some modalities are much more appropriate for certain categories of actions than for others.

To proceed further with our attempt to develop a framework for discussing behaviour, we would have to specify what is the nature of the acts which we have categorized as describing, manipulating and exploring. Of what does describing consist? In what circumstances would we say an organism is manipulating and in what circumstances, exploring? Before we can answer these questions we need to develop a more adequate structural description of the notion "situation". We shall attempt this task a little later while discussing language behaviour in greater detail. The richness and variety of language expressions would seem to be essential to talk about aspects of situations adequately.

But before we do this it may be useful to add a few more comments to complete our meta-level characterization of an organism as a behavioural system. Notice, first of all, that when we refer to "situations" in P3, it is the *total* situation we are referring to. A particular behavioural act of an organism could describe its own internal situation, that is, its own state. In the context of language behaviour, this description could be in the form of statements about the self. Similarly, manipulation could very well be manipulation (and/or control) of the self (i.e., the internal situation) and not merely aspects of the external environment. Examples of these are: speech as therapy, protocols in problem-solving, therapeutic play, and so on.

Next, it is clear that a complete meta-theory of behaviour must also take into account the dynamics of behaviour. There are two broad aspects to dynamics: (1) learning the behavioural schemata in terms of which situations could be described, manipulated and explored; and (2) using these learnt schemata appropriately to describe, manipulate and explore specific (given) situations. We must specify what the universals of learning mechanisms are and how they function. Learning essentially consists in: (1) learning new schemata, (2) learning new compositions of available schemata (i.e., learning new structures made up of constituents which are available structures), and (3) relating these to aspects of new situations encountered. Elsewhere we have discussed at length the current status of learning theories in psychology and their general inadequacies (see, Narasimhan, 1978).

We shall return to this problem of learning and make some specific comments when we discuss later the developmental stages of language learning. But, referring to universal principles of learning, we might note in passing, now, that learning from examples would really seem to be one of the primitive modes of learning. Analogizing based on learned paradigms, it is possible to argue, is the fundamental behavioural principle of transfer and generalization. An animal tends to extrapolate to a new situation behaviour it has learnt in an old situation to the extent to which the new situation bears a resemblance to the old one. The conditioning principle of psychology can readily be reinterpreted along these lines. Analogizing principles based on the concept of similarity would thus seem to play a fundamental role in behaviour modelling. Giving a computational interpretation to the notions "similarity" and "analogizing" is perhaps the most central open problem to be tackled in this context.

Language behaviour as a second-signalling system

We asserted earlier that the universals of behaviour are "describing", "manipulating", and "exploring". We shall now try to characterize the substantive nature of these behavioural categories with specific reference to language behaviour. We should then be in a position to exhibit the relationship between language behaviour and behaviour in other modalities, and discuss in what sense language expressions function as second-signals.

The pragmatics of language behaviour are concerned with delimited totality of situations which may be referred to as the *world* to which the (particular) language relates. A given language, in other words, relates to some given world. Natural language behaviour at the naive level relates to the situational aspects of the natural physical world available at the (unaided) input–output interfaces of human beings (we consider later the non-naive usage of language, for instance in scientific theory construction). We shall now consider in a somewhat formal manner the categories embedded in natural language in its naive mode of functioning.

Primitively given are sets of *agents, objects* and *actions*. Agents are distinguished from objects by the fact that actions are performed by agents, but not by objects. Every agent, thus, is characterized by an action repertoire that that agent is capable of performing.

Among the totality of actions two subclasses are distinguished and referred to, respectively, as computing (evaluating, measuring, finding, determining) values of attributes, and values of relations. Attributes are, primitively, partially computable functions defined over the sets of agents and objects, and assume values from well-defined ranges. Having a specified value in a particular attribute is called a *property* of the object or agent under consideration. As illustrations of these concepts consider the following.

Length is an attribute applicable to the object, *this stick*. Let *length* (*this stick*) = 5 inches. Then, *5 inches in length* is a property of *this stick*.

Using attributes, objects can be classified into object types as follows (analogous remarks apply to agents): objects 01 and 02 belong to the same type if and only if all attributes applicable to 01 are applicable to 02 and vice-versa. Notice that this classification into object types defines an equivalence relation. This classification should not be confused with other kinds of classifications of objects and agents: e.g., classification according to properties (all things red in colour); classification according to performable actions (all things that can walk); classification according to specified relationships (all things greater in length than that object); and so on.

The *state* of an object (or agent) is completely specified if the values of all its attributes are completely specified. Since the totality of attributes applicable to an object (or agent) may be a very long list (or even not completely known at any given time), it is customary to restrict attention to delimited subsets of attributes only at any given moment. Thus, normally, only *partial states* are specified. Let us use the term 'state' to refer to 'partial state', to simplify terminology, and use 'total state' to refer to 'state' whenever required.

Relations are partially computable predicates defined over two or more states of one or more objects or agents. Notice that, in particular, relations are defined over values of individual attributes. As an illustration consider this: *greater than* is a relation defined over values of the attribute *length*. Length is applicable to the objects, *this stick* and *that stick*. Hence, *length (this stick)* greater than *length (that stick)* is either true or false.

Agents perform actions. A consequence of an action is a change of state of one or more objects and/or agents. An *elementary event* is an agent in action, or the change of state of an object or agent. An *event* is a structured complex (i.e., a network) in time of elementary events.

A *situation* is a structured complex of agents, objects and events. The totality of situations constitute a *world.*

In terms of the structure that we have thus assigned to a situation, we can now discuss the nature of the universal behavioural categories, describing, manipulating and exploring.

Describing a situation consists in specifying what objects and/or agents are present (i.e., are aspects of the situation being described), what their states are, what relationships, if any, exist between these states, what events are happening (i.e., are aspects of the situation being described), what relationships, if any, exist between these events, and so on.

Manipulating consists in causing events to happen.

Exploring consists in determining what descriptions apply to the situation under consideration.

In the language modality, these meta-behavioural categories are identified by a variety of names. We list a sample of these below for each category.

Describing	describing, asserting, stating, informing, . . .
Manipulating	manipulating, commanding, controlling, ordering, permitting, requesting, persuading, instructing, coercing, . . .
Exploring	exploring, probing, querying, asking, demanding, verifying, evaluating, comparing, solving, puzzling-out, experimenting, . . .

Let us refer to a language expression that relates to (refers to) some specific aspect of a particular situation as a *language behaviour token* (lb-token). The unit of behaviour in the language modality is an *utterance.* An utterance is either a lb-token, or a language expression representing a language schema, or obtained from a language schema through the substitution of specific lb-tokens for the proforms occurring in the former. The utterances in a natural language describe, primitively, the states and relationships as computed by the sensory-motor mechanisms, and command and control actions performable by the motor mechanisms. Explorations, again, relate to aspects of situations arising in these contexts.

Natural language interaction between two human beings, then, relates, to discourse about, and manipulations of, some class of situations of the naive physical world. The interaction consists in making assertions about the states of objects and agents in a given situation and about the relationship between the states, commanding and controlling actions that cause events relating to the objects and agents in the situation, and asking questions about the objects and agents, their states, relationships between the states, and so on.

We can now elaborate our thesis concerning the second-signalling (i.e., mediating) role that language behaviour plays and how it plays this role. Our assertion is that in any language there exist only a limited number of different schemata of descriptions, commands, controls and questions. In studying a language we study what these schemata are and what classes of situations they relate to.

Learning or acquiring a language consists in acquiring (or, more precisely, building up) the schemata provided by the language to generate descriptions, commands, controls and questions, and in using them to describe, manipulate and explore particular situations. The *interpretational system* that gets built up as a child acquires linguistic competence consists of lb-tokens and their structured relationships in terms of associations between:

(a) lb-tokens among themselves,
(b) lb-tokens and sensory-schemata,
(c) lb-tokens and effector (motor) schemata,

(d) lb-tokens and state variables.

Thus, the problem of interpretation of descriptive utterances (i.e., the problem of relating them explicitly to the structured aspects of situations) consists in analyzing the utterances into lb-tokens and their inter-relationships and, thus, apprehending the situational aspects and their inter-relationships. Similarly, the interpretation problem for command/control, and query utterances consists in performing the appropriate action commanded in the first case, and generating the appropriate response in the second case. Analogous remarks apply to generation of utterances which are descriptions, commands/controls, or queries.

It is to be noted that an interpretational system as we have described it above is a characteristic of an individual. The details of the interpretational system of a given individual could be inferred only through careful studies of the individual's behaviour in a variety of well-circumscribed and well-articulated situations, and by trying to establish correspondences between the lb-tokens in the elicited behaviour, other observed non-linguistic aspects of behaviour, and the several situational aspects.

However, it is clear that the aspects of the situations as determined at the external (i.e., sensory-motor) interfaces should be comparable, by and large, among the individuals of a language community. Hence, to make communication through language possible at all, the correspondences between the lb-tokens and the situational aspects at the external interfaces should also be, by and large, comparable among all the individuals of a language community. Thus, among these members the details of their interpretational system should agree to a large extent. These are just the aspects that show up as the common aspects of the language behaviour of that community. This is just another way of saying that a child builds up its interpretational system by interacting with a language community and, *hence*, in conformity, by and large, with the interpretational systems of the individuals of that community.

Before concluding this sub-section, it may be worth pointing out that the language categories and the pragmatics of language behaviour that we have been considering so far can be extended in a natural way to language behaviour that relates to situations not part of the naive physical world. Artificial languages are, thus, those constructed to cope with (i.e., describe, manipulate and explore) artificially constructed worlds. In scientific theory construction, for example, these artificially constructed worlds are (considered to be) formalized abstractions of aspects of the phenomenology of the natural physical world. Such languages, in general, are constructed formally by defining sets of objects with attributes, and action schemata which determine attribute values of given objects, transform one set of objects into another, and so forth.

Artificial languages designed for computer work define worlds whose situations are problem situations. Describing and manipulating these situations are treated as problem-solving activities. In computer languages, as they are in use at present, in general manipulations play a much more dominant role than descriptions.

In story-telling what presumably one does is to use a natural language to construct descriptions of an artificial world, i.e., a complex of fictive or imaginary situations. We "people" them with agents and objects with fictitiously assigned attributes, states, and action repertoires. The affective connotations of the lb-tokens available in the interpretational systems of the story-teller and the listener, clearly, determine the affective status of the situations described in the story. These affective connotations mirror the associations, either direct or indirect, between the lb-tokens and the internal state variables.

The language acquisition process

So far we have been discussing the thesis that acquiring a language consists in acquiring certain classes of lb-tokens and schemata, and relating them among themselves and with the situational aspects in specific ways. We have not said anything up to this stage about the actual acquisition process. That is, we have not said anything about the kinds of primitive processes that must be postulated as being available to the child so that it can build up these specific schemata and associations by interacting with the language community.

From the point of learning theories as formulated and studied by psychologists, one could ask whether language acquisition could be accounted for by invoking only the classical and instrumental conditioning processes. Skinner takes the view that it is unnecessary to postulate any new kinds of processes to account for language learning. Chomsky, on the other hand, argues that language acquisition is a process that is *sui generis*, and that it is dependent on unspecified (unspecifiable?) innate endowments. Clearly, both are extreme and untenable views.

The classical and instrumental conditioning processes require careful and systematic tutoring involving the use of reinforcements explicitly administered by the tutor. The Gardners in teaching sign language behaviour to Washoe had to go through such involved and laborious training procedures. These, more or less, correspond to the Skinnerian paradigm. Children, however, acquire language with great ease and spontaneity requiring hardly any tutoring or reinforcement in the normally understood sense. It is, hence, difficult to agree with Skinner that the conditioning processes (as the psychologists currently understand them) are sufficient to account for child language acquisition. Nevertheless, in view of the well-established roles these processes play in animal learning, it would indeed be appropriate to assume that they do play a significant role. But additional processes have to be postulated which substitute for the role of tutors and deliberately administered reinforcements. The problem consists in determining what these processes can be. It is quite evident that many of these must form part of the built-in capabilities of the child considered as a behavioural system and play a central role in child development in general, and not merely in language acquisition.

It seems plausible to argue that imitation, rehearsal, and analogizing are three closely related and interdependent processes that form the behavioural substrate of the developing child and play a significant role in language acquisition. It is well-known to psychologists familiar with child development that children tend to imitate all aspects of the behaviour of others and this is especially apparent in motor and verbal areas.

Till recently there was a tendency in psycholinguistic studies to dismiss somewhat casually the importance of the role of imitation in language acquisition: this was based on the argument that the child's and the adult's utterances are not identical and, hence, the child cannot be said to be imitating the adult. Also it was assumed that learning through imitation could not teach the child how to be productive.

However, in the last 5 years or so, psycholinguists have increasingly begun to veer back to the classical view that a child's utterances cannot be related directly to an adult's at the purely verbal level, since the former is partially verbalized and partially "acted-out". These body-action schemata really signify the relational structure connecting the one or more words verbalized. The adult's utterance nevertheless serves as a *model* for schematization, and the child's behaviour is imitative in the sense that just this schema is associated with the apprehended situational aspects and is made use of again when the same or similar situational aspects are encountered. Imitation and analogizing together can be shown to be capable of realizing a productive process.*

In order to determine with much greater definiteness what kinds of underlying mechanisms and processes have to be postulated to account for child language acquisition, there is an urgent need to analyze and systematize from a computational viewpoint the vast amount of very diverse and rich field data relating to child language that is beginning to be accumulated.

Computation Modelling of Language Behaviour

Modelling language usage

We shall restrict our discussions here to language usage in con-

*R. Clark: What is the use of Imitation? *J. Child Language*, 4, 341–358, 1977.

versation at the naive level, since our primary interest is with naive language behaviour. Consider what is involved in conversation at the most elementary level. The conversing partners talk alternately, but there is an internal coherence to this alternating sequence of individual behaviour. If A and B are conversing and A makes a remark, he expects B to *respond* to it. An important technical issue here consists in being able to articulate what A's expectation really amounts to and how he verifies whether B's response is in consonance with his expectation. (Parenthetically one may note here that very early in the language acquisition process children engage in "Social Play" or "Pretend Play" the central aspect of which is "role-playing". The verificational ability that A displays above may be viewed as an instance of "role-playing".)

The response verification and acceptance takes place within a delimited context. The contextual clues form part of the background information that A has; these are the presuppositions that he brings to bear on the conversation. It must be noted that this contextual background is continuously being modified and augmented as the conversation proceeds. B's answer must not only be in keeping with A's presuppositions of the immediate context of the conversation but also with his general background knowledge about the type of world being discussed.

A conversation relates to some specific situation. The utterances relate to various aspects of this situation. As the conversation proceeds, the conversing partners accumulate more and more information about this specific situation under discussion. But this specific situation is an instance of a whole class of similar situations that could potentially occur. One's general knowledge of the world is built up of characteristics of these classes of situations. This knowledge is acquired through one's personal experience in interacting with the world, through one's studies, from stories, and from a variety of other sources. This knowledge serves as a framework within which the particulars of the specific situation under discussion are tested. These particulars are expected to be self-consistent as the conversation progresses.

Notice that it is precisely this background knowledge that

serves as a general framework enabling one to ask relevant questions to probe. One knows what details to expect and, hence, one knows what details are missing in a given account of a situation. It is not essential that all aspects of a situation have to be explicitly stated. It may be possible to deduce some of the missing details from other available details. Conversational language behaviour, thus, requires problem-solving competence at a naive level. One's conversational capability is also seen to be delimited implicitly by one's familiarity with the world of situations being discussed.

Earlier we pointed out that a child, as it acquires its language, builds up an interpretational system. This interpretational system is precisely the one that is involved in the understanding of utterances as we have been discussing this process so far. Clearly, then, in modelling conversational language behaviour the crux of the problem consists in modelling this interpretation assignment process. Computer scientists, of course, have always been aware of this in designing computational (i.e., programming) languages. The formalization of the interpretation assignment process in this context is accomplished in a certain standard manner as follows.

(1) First, one defines a set of primitives or atoms with assigned interpretations.
(2) Then, one gives rules for constructing larger language entities, i.e., expressions, phrases, statements, etc., out of these atoms and other already constructed entities. These rules, in general, could involve recursive procedures.
(3) One, then, defines an analysis procedure for *parsing* a given expression, or other higher level entity.
(4) The interpretation to be assigned to an expression (or some other higher level entity) is then defined by a certain set of *fixed processes* which are dependent on the parsed output as well as the interpreted atoms occurring in the expression (or higher level entity) in question.

It is easily seen that this way of structuring the interpretation assignment process assigns a natural structure to a computational language. One can refer to the atoms (vocabulary), the formation rules, and the parsing procedures as constituting the syntax of the language, and the interpretations assigned to the atoms (dictionary?) and the set of fixed processes for assigning interpretations to the higher level entities, as constituting the semantics of the language. It is also evident that syntactic considerations play an intrinsically fundamental role in this way of formalizing the interpretation assignment process.

A translator for a computational language effectively has to make use of a grammar of that language. Parsing procedures must be explicitly given to analyze any given input string. And these procedures must have incorporated in them the vocabulary and the rules of formation either explicitly or implicitly. In this sense, the interpretation assigning mechanism must have the grammar of the language built into it.

This is the view that computer scientists have adopted all along as regards interaction with computers in the natural language mode also. Perhaps the most impressive natural language interaction program that has been constructed up to now is the one due to Winograd*. This system is set up to cope with situations in a toy world consisting of coloured blocks of several shapes which can be configured in various ways. The system can move these blocks around as instructed, build new configurations, and talk about what it is doing and what it has done. This system has, possibly, one of the most completely articulated grammars of English built into it.

Linguists, as well as psychologists interested in natural language behaviour, would find it well worth their effort to study carefully and in detail Winograd's implementation. If they do so, they would find that the essential strength, as well as the principal weakness, of Winograd's system consists precisely in the fact that its linguistic capability is delimited explicitly by the grammar of English that has been built into it. For, this grammar defines not only how an input string is to be parsed but also how it is to be interpreted. But, as we discussed earlier in this section, how one

*T. Winograd, *Understanding Natural Language*, Edinburgh University Press, Edinburgh, 1972.

interprets an input statement (i.e., one's understanding of it) is delimited by one's knowledge of the world. In order to cope with new situational aspects, appropriate additional knowledge has to be supplied. For this knowledge to become effective in the interpretation assignment process, it has to be integrated with the existing knowledge.

In Winograd's case, new knowledge has to be added to the built-in grammar as the interaction proceeds and new (hitherto unencountered) situations arise. Since the system's grammar, as it stands now, has been completely worked out by Winograd and incorporated in the system through the use of a special programming language, there is no way of altering it (or augmenting it) except by reaching into the system and incorporating additional chunks of programs. Only Winograd can do this, or only someone who knows sufficiently well the inside of the existing system and the special programming language used in its implementation. In other words, although the system has ostensibly been designed to understand English, and has a rather massive and efficient grammar of English incorporated into it, there is no way of *teaching the system in English* new knowledge of the world.

The system, to be sure, can accept a certain limited set of definitional statements in English, but substantially our central criticism remains valid. Winograd himself points this out and remarks there is no simple way of augmenting or modifying its present behaviour repertoire through input statements in English. One cannot tell the system, for example, "when you want to pick up a block, first take everything off from the top of it", and expect the system to incorporate this as the initial segment of its "picking-up-a-block" procedure. Analogous weaknesses are to be found in the response generation part of the program.

In the usual context of using programming languages (in computing activity) such problems are not encountered because one is all the time dealing with a completely closed world of situations. The question of augmenting the interpretational capacity of the system does not arise. The programmer learns what the system can understand and writes his programs accordingly.

This, then, is the essential difference between interacting in a computational language, as computer scientists normally do, and interacting in a natural language, as children do. In the first case, the language, as well as the interaction, relates to a strictly closed world of situations. In the second case, the language, as well as the interaction, relates to an open-ended world of situations. It would be meaningless to argue that one could take into account all possible contingencies and build their implications into the grammar to begin with. In principle, one cannot deal with an open-ended world in this manner. Natural language behaviour must, hence, be necessarily based on teachability, teachability making use of that very language.

PLAS, a paradigmatic language acquisition system

We can summarize our criticisms of syntagmatic (i.e., explicitly grammar-based) approaches to behaviour modelling as follows: (1) there is no natural evaluation procedure for choosing between different grammatical frameworks as biological evolution would demand; (2) there is no way of growing such frameworks or modifying them step-by-step (incrementally) as developmental considerations would demand; (3) there is no way of modifying and/or augmenting such frameworks through experience (i.e., through "particulars").

To satisfy the above three criteria what we need is a framework for behaviour that can be grown (augmented/modified) step-by-step through examples (i.e., paradigms). As we saw earlier in discussing children's language acquisition, a learning principle that has this feature is analogizing based on learned paradigms. Behaviour acquired in one particular situation is extrapolated to a new particular situation to the extent to which the new situation resembles the old one. The primary computation schemes in the paradigmatic approach to behaviour modelling are analogizing principles based on the concept of similarity*.

To test the viability of the paradigmatic approach, a

*R. Narasimhan: Modelling Behaviour: the Need for a Computerised Approach, *J. Social Biol. Struct.* **1**, 79–94, 1978.

preliminary version of a system, PLAS, that is capable of being taught through examples to function in the language modality has been implemented in TIFR, Bombay*. The language behaviour that PLAS tries to simulate is somewhat similar to the language understanding and usage behaviour that a student exhibits while going through a "self-teaching" language manual, say. The manual is typically broken up into "lessons", each lesson being concerned with a limited set of linguistic concepts and rules which are taught through illustrative examples. Every lesson usually ends in a set of "exercises" or "tests" which help to assess whether the student has been able to absorb successfully the fundamentals of that lesson. See for example Fig. 15.1.

It is worth explicitly noting some points about the class of task environments illustrated in Fig. 15.1. A lesson is a coherent text consisting of several inter-related sentences which together describe a complex situation; often such a text is supported by a visual illustration. "Understanding" the text amounts to understanding what this complex situation is, i.e., what its several aspects are and how they are inter-related. The questions probe the student's understanding of the text, primarily. At a secondary level, in answering the questions the student also exhibits his linguistic command over the language. For, understanding a question requires sentence comprehension capability, and answering a question requires sentence generation capability. The generated sentence must be grammatically well-formed and also be factually correct (relative to the complex situation described in the text). The "sentence completion" and "sentence construction" exercises, on the other hand, test the creative, language-using capability of the student. The answers to these exercises are essentially open-ended and are not delimited by the situational aspects discussed in the lesson. However, these answers cannot be arbitrary. They must relate to potentially realizable situational aspects (even if not to the specific ones discussed in the lesson).

*V. SEMBUGAMOORTHY: *PLAS, a Paradigmatic Language Acquisition System*, Ph.D. thesis, Indian Institute of Technology, Bombay, 1978.

The essential design feature of PLAS is that it is teachable and that it functions in the paradigmatic mode. No part of a grammar of English is built into the system. Nor does the teaching mode involve teaching the system particular *grammatical* facts about English in a *formal* manner. The system does not construct for itself a grammar of English (in the linguist's sense) from the examples to which it is exposed in the teaching mode. What the system actually does is to analyze (on the basis of its currently available language behaviour competence) a complex situation into plausible aspects by converting it into constant parts (schemata) and variable parts (fillers or variables). It is capable of relating newly analyzed aspects of a situation to available, already-analyzed situational aspects. It is capable of synthesizing answers to questions about this situation, or constructing new sentences "creatively". The system accomplishes all these essentially operating through analogizing principles based on similarity considerations. Fig. 15.2 illustrates some aspects of schematization and similarity grouping that underlie the functioning of PLAS. Finally, Fig. 15.3 gives some idea of the kind of texts the current version of PLAS can cope with. It would take us too far outside the scope of this paper to go into the details of the behavioural mechanisms of PLAS. Sembugamoorthy's thesis should be consulted for this purpose.

The present implementation of PLAS, although quite rudimentary, clearly demonstrates the viability of the paradigmatic approach to language behaviour modelling. By incorporating in an essential way, as part of the structure of the model, analogues of the verbal environments and feedback processes that children are exposed to in the early stages of language acquisition, the paradigmatic approach enables a systematic study of the extent and manner in which these environmental features affect the developmental process. For example, before one can come to grips with the modelling of the understanding of *events* (as opposed to static world views which is all that PLAS is capable of handling at present) we would have to learn to model the sensory-motor space more adequately. What kinds of primitive action schemes and primitive integrating processes should we postulate so that using these more complex action schemes can be built up paradigmatically? How can more complex action schemes be built up through instructions (i.e., description) in the language modality? What additional primitive machinery is required to interpret language expressions belonging to the time modality? It seems plausible to argue that primitively the time dimension should be interpreted in terms of the system's own ongoing activities. Terms such as "before", "after", "again", "first", "next", "now", should be capable of being interpreted in this way to begin with. It would then seem to be possible to teach the system how to assign interpretations to other time expressions and schemata involving tense-markers.

The pragmatics of language usage in talking about one's own behaviour (or, equivalently, that of a conversational partner) raise several important issues. In modelling such usage of language it would seem to be essential to incorporate in the model "motivational states" or internal control parameters that essentially play this role. In order to be able to interpret and act appropriately on the basis of statements of the type, "When you are in doubt/difficulty do x", or "If you want to accomplish y, do z", etc., the system must know *when* it is in doubt/difficulty, or *when* it is trying to accomplish something, etc.

The second-signalling role of language behaviour can be appreciated most clearly when one considers the methodological aspects of teaching an organism to do something when in some internal state. In the absence of language, this teaching can be carried out only when the organism is in the corresponding state. The organism must be experimentally prepared to be in that particular state, and then be made to carry out the specific act so that the desired outcome results. This is the well-known operant conditioning technique. With the availability of language, on the other hand, instructions could be imparted by referring to the specific states by name without requiring that the organism *be* in that very state at the time of training. It is only necessary for the organism to be aware of the names of the states it is in so that the appropriate instructions could be put into effect by relating together the names and the instructions. It is this interface

A. Task Environment

Ability to be taught situational aspects described by a "text" and to demonstrate the behavioural competence acquired through (i) answering questions about the text; (ii) understanding a similar new text; (iii) constructing sentences using given language expressions.

Lesson

This is a pencil. This is John's pencil. It is on his desk. It is red. John's book is also on his desk. The book is not red. It is green. It is also heavy. It is heavy because it is large. The girl next to John is Mary. Mary's books are on her desk. Her notebook is inside her desk. There are many class rooms in the school.

Exercises

1. Questions
   a) What is the colour of John's pencil?
   b) Where is it?
   c) What is on John's desk?
   d) Why is his book heavy?
   e) Where is Mary?
   f) What are on Mary's desk?
   g) Where are Mary's books?
   h) What is inside Mary's desk?
   i) How many class rooms are there in the school?

2. Complete the following sentences
   a) Mary is ____________
   b) Many class rooms are ____________
   c) John's pencil is ____________

3. Use in sentences of your own
   a) (Peter, pencil, table)
   b) (Pen, box)
   c) (Peter, Jane)

Fig. 15.1 PLAS, A Paradigmatic Language Acquisition System (V. Sembugamoorthy)

between language behaviour and motivational states that needs to be mirrored in the model.

Yet another crucial aspect of modelling that requires careful study concerns the role of language in the enrichment of the teaching mode. This is the central aspect of the use of natural language as a meta-language. In the present version of PLAS the teaching methodology remains fixed and pre-wired, as it were.

Language acquisition is predicated on the availability of sensory-motor structures in terms of which the situational aspects of the world can be apprehended.

The interpretational system that gets built up during language acquisition consists of language-behaviour tokens (lb-tokens) and their structured inter-relationships through associations between

(i) lb-tokens themselves via language schemas
(ii) lb-tokens and sensory-motor schemas
(iii) lb-tokens and sensory-motor tokens.

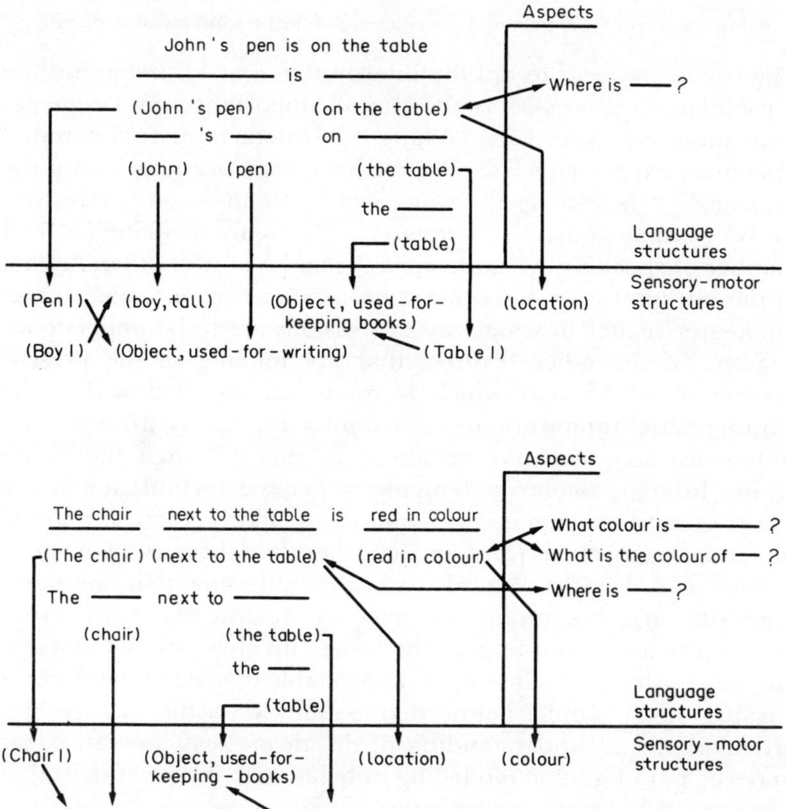

Fig. 15.2 PLAS: Fundamental Principles

Having been taught "Text 1" in the teaching mode, PLAS could understand "Text 2" in the testing mode.

Text 1: A chair and a table are near the window. The chair is brown in colour. The table is red in colour. A desk and a chair are in the middle of the room. The brown chair is made of wood. The one in the middle of the room is made of plastic. The chair made of wood is heavy.

Text 2: This is a chair. There is a box on the chair. It is heavy. The box inside the shelf is light. The heavy box is green in colour.

Fig. 15.3 An illustration of the competence of the present version of PLAS

The use of *comments* to aid the interpretation assignment and the consolidation processes is of utmost importance in language-acquisition and interpersonal language interaction. In a running discourse, expressions like, "as we have been arguing", "on the contrary", "in exactly the same way", "and so on", "respectively", "analogously", "similarly", "roughly speaking", "in a manner of speaking", "so to speak", and a host of others, relate to the interpretation assignment process and are intended to tell the hearer/reader in what way expressions are to be understood.

Some of the other features that are missing in the present version of PLAS and which have to be modelled within the paradigmatic framework are: naive inferring (going from general statements to particular instances of them), naive theorizing (going from particular statements to general formulations), the use of quantifiers (some, all, none, a few, many, etc.), the use of sentential connectives (and, or, not, therefore, because, for, etc.).

After two decades of hard work, it is only now that computer scientists are beginning to have a reasonably satisfactory theoretical understanding of the issues involved in syntagmatic language processing. It is not unreasonable to predict that a comparable effort would be needed before we achieve a similar articulation and understanding of the theoretical problems that underlie paradigmatic modelling of behaviour, in general, and of language behaviour, in particular.

CHAPTER 16

Fundamental Concepts of Eco-technology

S. AIDA

Development and Ecology

In much modern scientific and technological development, nature (the "natural order of things") has been regarded as an opponent to be conquered rather than as an asset to be utilised. Therefore, much of the history of civilisation has been a history of struggle, originating in the determination of the human race to survive.

Nature as an object to be developed and used is generally described as "resources". To date, man has concentrated most of his attention on how to make good and efficient use of these resources, and nearly all technological progress has been dependent on such use. Thus, technological development has led to the gradual expansion of all forms of geological exploration, now entering probably its last phase on earth with the exploration, and exploitation, of the oceans.

Along with the quantitative increase in the practical uses of science and technology, related environmental problems are now receiving ever increasing attention, with severe criticism of any development (such as off-shore oil wells) that does not pay maximum respect to the protection of the environment. A new wave of opposition to environmental damage has been stimulated by events such as the pollution of the beautiful Californian coastline following an accident at an off-shore oil rig near Santa Barbara, or the pollution of the Brittany coast of France following the stranding and break-up of a large oil tanker on passage up the English Channel.

Quite independently, in fact, of warnings such as those of the Club of Rome, the increasingly rapid (some would say profligate) exploitation by modern society of limited natural resources, largely aided by recently developed scientific and technological methods, is presenting mankind with a new and formidable problem: that of preserving the *milieu* in which he lives—and on which he is dependent for his life.

Defeatist reactions to scientific and technological developments have often been stimulated by their clumsy application. Some have claimed that technology is "running wild", and must be restrained before it threatens the survival of the human race. Such attitudes deny the fact that science and technology are "neutral", and whether they bring benefit or disaster depends on how they are used: the argument that they are intrinsically dangerous or evil is nothing more than an evolution of the ethical doctrine that man himself is intrinsically evil.

There are two approaches to establishing the pursuit of scientific and technological development as a valid exercise. The first—and it may seem to be rather paradoxical—is to maintain that no validation is called for: much development originated simply from more or less haphazard curiosity and enquiry by dilettanti.

The second approach considers development in relation to its social role, and attempts to rediscover the very starting point of development itself. This does not mean (as is sometimes claimed) a "return to the starting point" of all our thought processes, but rather that we should re-examine the origins of those thought processes in order to establish a new set of values by which all development should be judged, and by which future development should be directed. The basis for these new values must surely be "ecology" as an all-embracing science; their objective the confirmation that all modern scientific technology, in the biological world around us, must be in harmony with, and a component of, nature. This calls for a profound reassessment of the reasoning behind development, in particular the abandonment of the "confrontational aspect'' of science and technology.

In the biological world, life is maintained by many multi-lateral and interdependent systems, all contributing control functions to preserve overall stability. For the activity of all living things, energy is necessary. Its basic source is the sun*. Trees, plants and vegetables, under the influence of sunlight, can transform by photosynthesis some of the energy from that sunlight into the chemical energy necessary for animals to live. Without this essential process animal and human life, in the form that we know it on earth, could not continue.

Despite this clear and evident fact, when we begin to study the reasoning behind scientific and technological development, we find that much of it so far has ignored the need to maintain harmony with nature. Instead, the main emphasis has persistently been on demonstrating the predominance of physical science, often to the detriment of the ecological viewpoint that biological stability is central to survival. While it can be admitted that this is not the fault of development itself, it must also be accepted that there have been incorrect interpretations of the role of science—leading to a mistaken evolution of technological methodologies. The author believes that the modern science of physics is particularly vulnerable to this charge: few, if any, developments of modern physics have been evolved or regarded as starting points for human effort to preserve the biological world. Instead, the science of physics has all too often been oriented towards self-preservation and continuation of short-sighted scientific technology.

It is now vitally necessary for all of us, from the standpoint of preserving human society and civilisation, to establish efficient methods for the synthesis of artificial products to supplement (and in some cases to replace) those natural ones which we are

*It is of course true that much of the energy now being used to support an ever-increasing world population is energy that came from the sun many millions of years ago, being now stored in the form of fossil fuels which are clearly insufficient to meet demand for ever. Even non-fossil fuels (such as uranium, thorium, hydrogen, lithium) must owe their initial existence to the peculiar set of circumstances which led to the formation of our sun and our solar system.

consuming too rapidly. We must also find and pursue a development philosophy which recognises that scientific technology cannot continue to advance other than as part of the dynamic continuance of the ingenious combination and harmony that is the biological world.

In such a way we may hope to rid our development philosophy of its former centralisation and concentration on physical science, and to create for it a rationale which fully embraces the doctrines of ecology. Thus may we understand correctly the present situation and future role of scientific technology, and its significance within the overall impact of all development.

Systematic Research

The original and basic meaning of *Ecology** is that branch of biology dealing with the mutual relationships between living organisms and their environment. More recently, the term has tended to become associated with the protection of the environment against real or postulated damage caused by human activities.

In Japan, ecology has been translated as ''seitai-gaku'', meaning the biology of living things and especially their relations with their environment. As such, ecology has evolved into a very special domain, becoming in fact a subject for basic contemplation in order to develop new philosophical principles of science and technology, leading to a new combination of every existing science in a single, harmonious whole.

Looking at the evolution of ecology, we can thus conclude that today it is concerned to develop, with every possible help from human genius and intelligence, an environmental system in which all living things, including human beings, animals, plants etc., can exist together—and continue to exist—in the surrounding material world comprising their environment. That means that ecological considerations today are based on dynamic rather

*The word (derived from the Greek ''oikos'' and ''logos'') was first used by the German naturalist and philosopher Ernst Heinrich Haekel (1834–1919) to signify ''the logical relationships in mankind's home and economy''.

than static analyses and beliefs, on evolving processes and values rather than on absolutes.

Viewed in this way, ecology may be interpreted variously as follows:

(1) as the underlying theory of a dynamic environmental system, and an all-embracing science susceptible to the influence of human beings;
(2) as a theoretical system-model comprising a multiplicity of closed cycles in a multitude of inter-related domains;
(3) as a theoretical method—based on advancing technical understanding—of enquiry into environmental unknowns;
(4) as a philosophical principle of technology, with the premise that mankind is, and should be, the central figure.

Thus the applicable fields of ecological study extend over virtually all branches of science and technology where human beings may intervene. Starting from the basic ecology of the atmosphere, forests, plains, rivers and lakes, seas and oceans, which are all essentially natural phenomena, the applied fields must deal extensively with the impacts of animals and human beings—their individual behaviours and the effects of their communities—in the case of human beings the substantial effects of towns and cities, transport, agriculture etc. Among the most important objectives are the elaboration of analytical methods freely convertible from the individual to the whole and from the whole to the individual. This should finally lead to a systematic theory for an environmental structure unifying element-oriented and system-oriented sub-structures in a single whole. In other words it should lead to the creation, on a basis of ecological concepts, of a scientific technology which closes the gap between micro-ecology and macro-ecology.

Because the above is becoming more and more widely accepted, ecological considerations have in the recent past acquired a rather strong influence. However, there has been a general tendency to examine these considerations mainly (if not only) in the framework of environmental modifications leading to problems of pollution. Ecology has thus become a major factor in criticism of the modern industrial system, and it is a matter for regret that this has sometimes impeded the inclusion of ecological provisions in evolving scientific technology.

In this situation it must surely be seen that there is a need to introduce into every individual branch of science a generalised critical appraisal from the ecological viewpoint (Fig. 16.1). But it should also be recognised that if the appraisal is strictly confined to criticism, and allows no room for productive *collaboration* between technology and ecology, the essential matter of the problem will be missed.

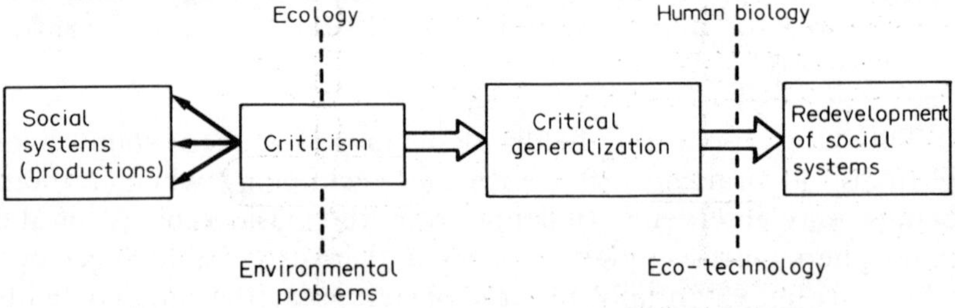

Fig. 16.1 Progress from ecology to eco-technology

Philosophical Principles of Eco-technology

Various approaches have been suggested to the problems of environmental protection in the face of massive consumption of material resources. Most are based on some form of resources circulation. For ecological concepts to be included, they must clearly form part of a new "all-round and multi-level" scientific philosophy, resembling closely that of the ancient Greeks in the early days of scientific technology.

Such a "humanised approach", far from the arrogance found in today's technology*, would be realised with the humility and respect for nature of those ancestral times.

When we consider the history of technology (which is a combination of pure science and technical art) we can trace it back to an original philosophical principle. The process of development from that principle to present-day engineering is illustrated in Fig. 16.2, where it will be seen that *art* is shown as a step between philosophy and technology, in the evolution of methodology to express an abstract philosophical principle in a concrete form. Leonardo da Vinci, whose interests were extremely diverse, was a representative of art in this sense, putting forward at an early stage ideas which can now be recognised as essential to the subse-

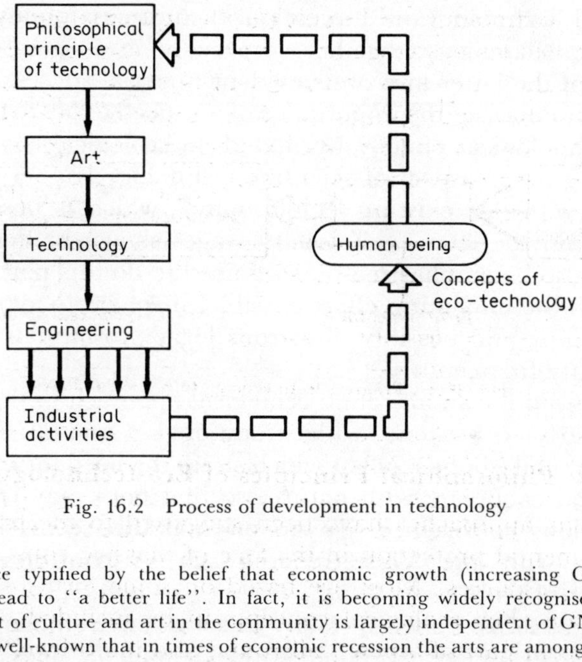

Fig. 16.2 Process of development in technology

*Arrogance typified by the belief that economic growth (increasing GNP) must necessarily lead to "a better life". In fact, it is becoming widely recognised that the advancement of culture and art in the community is largely independent of GNP (though it is equally well-known that in times of economic recession the arts are among the first to suffer).

The economic indicator of the "production first" policy has very often been used to assess a project from various human standpoints. This is quite natural and easily understood from an ecological point of view. It should nevertheless be recognised that GNP has generally been pursued on a very one-sided basis—that of such fields as inorganic physics and mathematics—while the most essential organic phenomena of life have been neglected.

quent evolution of Eco-technology. Fig. 16.2 shows how technology first appeared, leading via engineering to a variety of industrial activities which are today stimulating new human reactions of a philosophical nature: reactions which may be described as Eco-technological concepts.

Now, looking at modern industry, we find that the greater part has been born directly into the engineering stage of Fig. 16.2, i.e. subsequently to the combination of science and art into technology. This is particularly the case in Japan, where most scientific technology appears to lack the contribution of art to the establishment of a philosophical basis. Many people today therefore find it strange (and unnecessary) to discuss philosophy and scientific technology together, especially at a time of strong growth of the latter and of its evident material success.

This emphasises the importance of our need today to establish a new technological philosophy, based on ecological concepts and involving every aspect of scientific technology, rather than continuing with our existing ''philosophy'' which is based almost entirely on the later evolutionary stages shown in Fig. 16.2. It should also be emphasised that we need to do this now, urgently and with the maximum effort possible, if we are to avoid leaving an unhappy and possibly disastrous legacy to human society in the twenty-first century.

This need for an all-round ecological approach to the future is particularly consistent with traditional Oriental (especially Chinese) philosophies, all of which stress the need for mankind to unite and cooperate with nature, so that both may continue in harmonious coexistence. To an oriental this seems in distinct contrast with the Western approach of confrontation and conquest of nature.

Relationships between man and his environment have of course preoccupied philosophers throughout the ages, though it has been virtually universal to assume man's intellectual superiority over other living things. Thus Aristotle maintained that the human being was a ''social animal'', who could tell right from wrong because he possessed language while beasts did not, and that this distinguishing ability was responsible for his establishment of social entities such as nations.

Aristotle was by no means alone. About one hundred years earlier the Chinese mystic Hsün-tzu had evolved the same theory, stating that trees and plants had no perception; that animals, though they had a perceptive ability, had no sense of judgment between right and wrong; but that human beings had both perception and judgment, which enabled them to lead a social existence, and to form groups with activities based on logical judgments. If these judgments were not correct, the usual result was war, leading to the break-up of the group. From this, Hsün-tzu deduced the need for such ethical codes as filial piety, brotherhood, obedience, etc.

Hsün-tzu preached the so-called ethical doctrine that man is born evil, while another Chinese philosopher Meng-tzu (Mencius) advocated the ethical doctrine that man is born good, but both doctrines are thought to have derived from Confucianism, to have comprised basically Chinese thinking, and to have been among the many theories developed within the Chinese philosophy of the period.

All persons capable of logical judgment must necessarily reach conclusions which will depend on their initial information—the starting point of their thinking. Most people, even though part of a societal group, at least begin their thinking life with their own existence as the central and most important item of initial information: furthermore their thought processes have evolved round the concept of man as an independent individual. Hsün-tzu said "Man has ambition by nature. If his ambition is thwarted, he will redouble his efforts to achieve it. His limitless desire will cause further strife. Strife will cause disorder. Therefore, it is necessary to establish a code of courtesy, to still his desire and at the same time apply suitable restraints, trying always to maintain harmony between the two".

This was Hsün-tzu's concept of how the contradiction between individual freedom and life in a collective group should be resolved. It is certainly derived from the Confucian concept of "Jen" or benevolence as an ethical basis to establish harmony. The practice of Jen means that all people should suppress their limitless desires and behave in accordance with "Rei" (propriety), that is, with a universal harmony within themselves and towards others. Confucius asserts that in this way it is possible to maintain order in social life and to establish a peaceful environment. Jen and Rei, in fact, embrace all the moral qualities of the true man of intellect and reason: loyalty, reciprocity, dutifulness, filial and fraternal affection, courtesy, good faith and friendship.

In the confusion and disorder of so much of today's world, both material and ethical, it is certain that renewed and strenuous efforts are needed to harmonise all the components of our environment, an environment which now includes far more problems and in far greater variety than existed in Confucian times. Yet the basis on which the sought-after harmony may be achieved may still be found in these tenets of ancient Chinese thought.

Today's world is a world of technology, ever more dependent on the results of scientific research. And with the advancing means and skills ("Know-how") of today, the practical applications of scientific results are being pursued more vigorously than ever before. Yet this pursuit is almost exclusively materialistic, neglecting what might be called "the spirit" in science and giving excessive importance to material and economic growth. As a result, scientific technology itself is being widely questioned and criticised, notably as a principal cause of environmental pollution, which has become a symbolic phenomenon for attack.

There are various kinds of such pollution which, though all termed "environmental pollution", in fact fall into two distinct categories: "Hard pollution" by physical chemistry, and mental pollution, affecting the human mind.

Most of the pollution reported in the press is of the first variety, but it is probable that the most critical pollution, not only for the individual but also for future society, will be found in the second (mental) category. For, while it is possible to invoke scientific technology to halt the advance of hard pollution related to physical chemistry, so far as mental pollution is concerned, in the first place it is much more difficult to recognise, and secondly much more difficult to counter.

As a first step to recognising the features of mental pollution we must "study the human being himself". But even this presents many difficulties, for it demands a new technological (ecological) philosophy, created as a complement—and in a sense as an antidote—to existing scientific technology, which seems to have achieved a capacity for spontaneous self-generation and self-nourishment that has taken it beyond human control. The characteristics of the required new technology, or Eco-technology, are illustrated by Fig. 16.3.

A most important and novel characteristic of the system shown in Fig. 16.3 is that, in addition to the direct feedback from mental problems leading to stimulation of new ambitions, there are

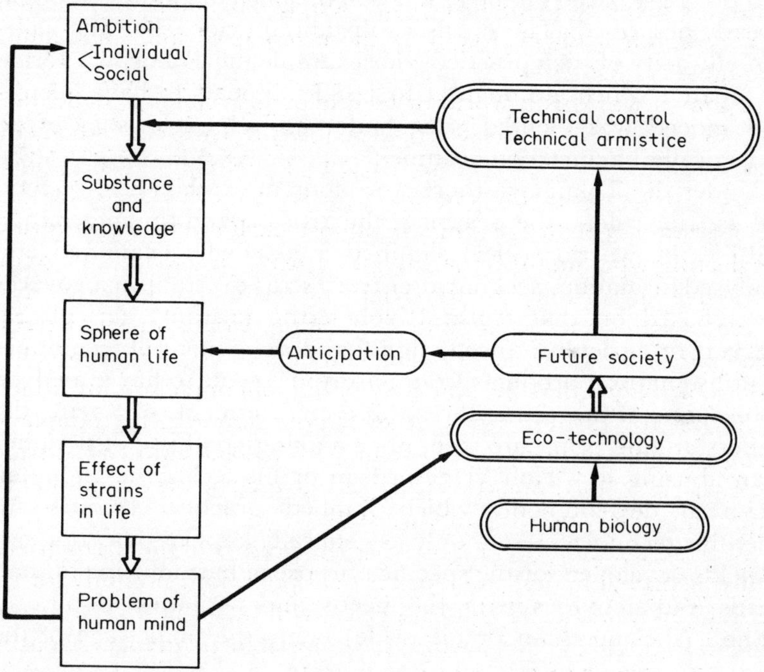

Fig. 16.3 Eco-technological system

further feedback paths via Eco-technology (which, as already indicated, derives its principal input from human biology) into different stages of the chain leading from ambition, through life and life's difficulties (strains) to further problems of the mind and hence further ambitions.

There are of course two kinds of ambition, individual ambition and that of a social group, each of which can create substance or knowledge (or both) and compose a "sphere of human life" which can be perturbed by the entry of "pollutions". The introduction of Eco-technology, to help resolve the resulting mental problems and confusions, should also lead to increasing harmony between the two types of ambition.

Taking an overall view of the system in Fig. 16.3, it is clearly and precisely based on the practice of the ancient Chinese thought already described. There exists substantial evidence of the value and efficacity of such practice—for example the Tokugawa period in Japan*, which although at first sight appears to have been—and indeed was—feudal, also had many characteristics of an ecologically balanced environment.

Under the Tokugawa there were four main classes in society: the warrior rulers, the peasants, the artisans and the merchants. "Shi" (the warrior ruler) signified a man who examined and analysed his nation, seeking after truth so as to establish a government based on that truth. "Noh" (the peasant, agricultural worker, or agriculture itself) signified development of techniques to utilise natural products from nature: it seems to have involved two aspects of protecting the natural environment, one being the understanding of nature, the other making good use of it rather than abusing it. "Koh" (the artisan or his technique) signified technical development, which implied practical recourse to scientific technology—not only as generally known today but also in fields developed for the specific purpose of harmonising human needs and activities with the needs and activities of nature. "Sho" (the merchant or his trade) being the final stage of the

*Latter part of the Japanese era of military rule (dictatorship) which began in the late twelfth century (Minamoto period) and lasted until the late nineteenth century.

system signified the pursuit of material benefit from nature both individually and collectively.

Comparing this system with today's society, there has clearly been a substantial change in the balance between the four classes: in particular the two last (artisanal technique and trade) have become intimately interconnected and thus acquired far greater influence than either of the first two classes alone—or even together. In fact, the order of the class system has reversed itself completely.

It is for consideration whether this may not be at the root of the strains and stresses of today's "modern society".

Basic Structure of Eco-technology

A simple definition of Eco-technology is scientific technology based on ecology. At first sight this appears clear and self-explanatory, but in fact it involves further definition of its component terms when used in this particular (holistic) conjunction. Before examining this, some other definitions are called for.

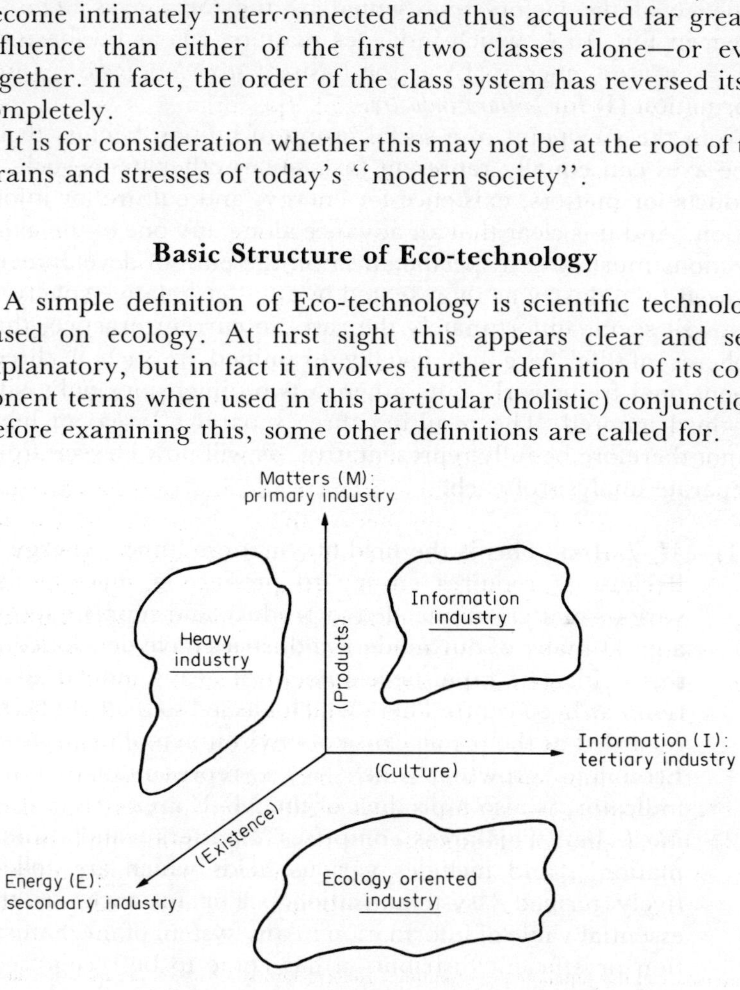

Fig. 16.4 Spaces of matters, energy and information

Modern civilisation can be regarded as comprising three main factors: matters, energy and information. If these are represented in a three-dimensional framework, as in Fig. 16.4 one can locate a "sphere of human life" (Fig. 16.3) somewhere in that framework, its position and *nature* depending on the relative importance or influence of the factors represented by the three axes. This is shown in Fig. 16.4, which indicates matters (M) as the basis of *primary industry*, energy (E) as the basis of *secondary industry*, and information (I) for *tertiary industry*.

From the viewpoint of a social group of human beings, these three axes can equally represent (as shown) other items such as products for matters, existence for energy, and culture for information. And it is clear that an advance along any one of the axial directions must have a specific effect on the path of development followed by industry as represented in any "co-relative sphere".

Now it seems unfortunately the case, in current practice, that "spheres of life" are not usually examined in such a three-dimensional framework, but rather in two dimensions only with the third ignored. The resulting three separate "areas of life" cannot therefore be fully representative, as will now be seen from a separate analysis of each.

(1) *M–E Area*. This is the field of "matters" and "energy". Because it requires energy to produce a matter, also various matters are needed to produce and supply energy, a great many of our modern industries have developed in this field. The rapid development of heavy-industrialisation has been centred here which has led to the field being regarded as the prime cause of environmental disruption. Economic growth, which is a representative GNP indicator, is also a product of the M–E area.

(2) *M–I Area.* This areas comprises "matters" and "information", and includes various fields which are collectively termed "systematisation". For instance, as the essential value of information in any system of mechanisation or efficient distribution has come to be recognised, and appropriate techniques have been developed,

systematisation (based on computers) has rapidly expanded and continues to do so.

As compared with the M–E area, which was mainly concerned with heavy-industrialisation, a major aspect of the M–I area is the "soft technique" of thought processes. As indicated by Fig. 16.4, the greater the influence of information, the more changes must result in the conception of things bordered by the material axis. This is a characteristic of tertiary industry, and many fields such as agriculture, forestry and fisheries, must be included in the area from this point of view. The area must also be regarded as the origin of *Cybernetics* as defined by N. Wiener*.

(3) *E–I Area.* This area is almost undeveloped in comparison with the former two. That is, energy and information have been developed individually through cooperative operations with materials, but the development of a field joining these together has been almost completely neglected up to the present time.

This is a field having close relations with some of the most vital phenomena in various levels of a modern "humanised system". Thus the neglect of this field, which may be termed that of *Eco-Energy*, is perhaps the weakest point of today's society. Its exploration and analysis offers basic knowledge necessary for the development of artificial intelligence.

We can see from the foregoing (and from Fig. 16.4) that any truly ecological approach to societal problems must involve a "space doctrine" of matter, energy, and information. This means that the development of a field belonging to the E–I area, with capacity for cybernetic control at various levels, is indispensable. A limited system of energy and information, in which materials will naturally intervene, can only play an inter-

*The American mathematician Norbert Wiener (1894–1964) invented the word *Cybernetics* (from the Greek term for "steersman") to describe the theory of communications and control mechanisms in machines and living things.

mediate role since energy, at any level, will change its form under the impact of information, to find its place in the area concerned. Clarification of this process should make it possible to establish a new system of technology effectively applicable in any field.

It would appear that, when it becomes possible to make specific observations of matter, energy and information on this new basis, the path from conventional technology to "Eco-technology" will become clear. Fig. 16.5 shows the process.

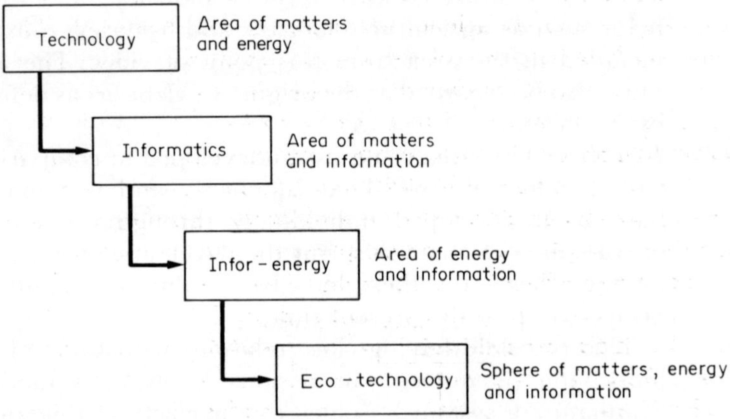

Fig. 16.5 Process from conventional technology to eco-technology

Next, to arrive at an appropriate thought process for the evolution of an ecologically-based scientific technology, three factors must be pursued: *structure*, energy, and information. That is to say, in the three-dimensional representation we have considered, material must be replaced by structure, and a new technological system for concrete development may then evolve.

The thought processes underlying such a structure in "dynamic space" were originated by Leonardo da Vinci, in his development of an anatomical chart. In this, he compared a human body to a single three-dimensional structure, and conducted all his research three-dimensionally in order to gain a physical idea of the arrangement of the inside of the body under the skin. This was the forerunner of our present methods in anatomical research.

Anatomy, however, is solely concerned with the physical structure of living things, studying how energy and information are combined dynamically in such a systematic structure, and the resulting effects on the dynamic actions of the structure itself. From this has evolved physiology, which now plays an important role in understanding the detailed biological functioning of human beings. Beyond both anatomy and physiology, we come to ecology as a dynamic science of the entire environment. This is the most advanced fruit of da Vinci's originality of idea in the field of scientific technology.

For a concrete group of three basic factors, Fig. 16.6 shows a "development space" in a 3-axis presentation similar to that in Fig. 16.4. This is not only helpful for understanding development space at various levels, and for planning a technological system evolved in that space, but it also presents a basis for the *theory* of

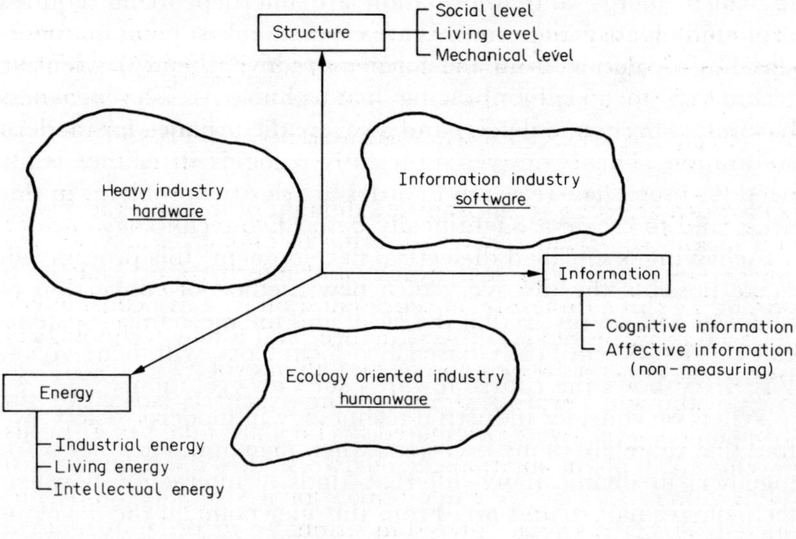

Fig. 16.6 Development space of structure, energy and information

ecological development. Moreover, it reconstructs an intellectual system as a field of learning, opening a thought process towards a new development theory.

This paper has already made much reference to energy, and its relationships with nature. When considering energy together with structure, however, the concept of energy can become confused and ill-defined. This is because too much attention tends to be given to heavy-industrial energy, with insufficient attention (or none) to the "moving energy of thought"—the energy of ecological living things.

When something moves, there is naturally a flow and transformation of energy, which for inanimate objects can be rather precisely analysed. But with human beings, flow patterns of energy are so different physically, biologically, and even spiritually that there is great lack of understanding, and a great need for research. Yet consideration of energy in this sense is extremely important in order to understand the significance of the structural space represented in Fig. 16.6. In particular, the area in which energy and information are interdependent requires profound investigation, for this area is the weakest point in understanding evolution from the former ("conventional") scientific technology to an all-embracing Eco-technology. The weakness has in fact become a defect, and a severe disturbance for modern civilisation. This is not yet sufficiently recognised; neither is the need for much new research in order to solve the problems in this area, and to create a scientifically-based Eco-technology.

Achieving a so-called dialectical development, this process will make possible the discovery of a new method of conversion of material and energy so that the *bionic* and the mechanical systems can coexist. Scientifically-based Eco-technology will then evolve. Fig. 16.5 shows the course towards such an evolution.

When we consider industrial technology in modern society, we find that there are many occasions when men and machines work together, involving many different kinds of interaction between technology, nature and art. From this viewpoint, if the development towards Eco-technology shown in Fig. 16.5 is postulated as part of an overall development within the global system of society,

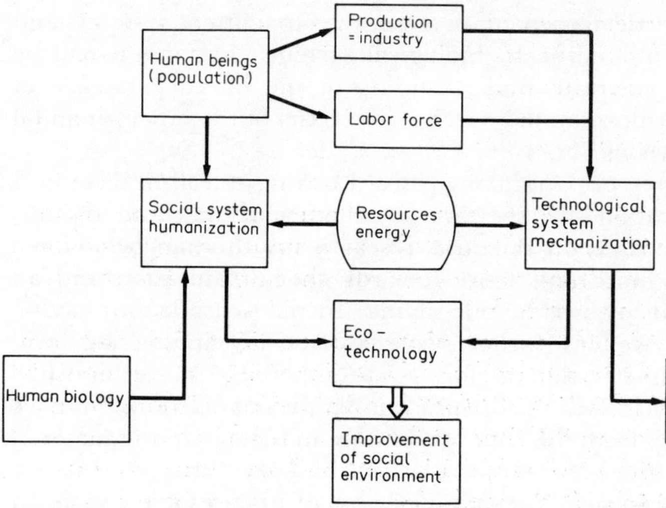

Fig. 16.7 Present environmental system and future

the complete process may be represented by a block diagram as in Fig. 16.7. This shows the parallel development, by the actions of the human population, of a materialist society at the same time as a "high-level" civilisation. It shows in particular how an increase of population having such a high-level material civilisation can cause industry to advance still further—although with little increase in the labour force if that industrial advance, being highly competitive, is achieved as a result of intensified mechanisation and automation. Technical revolution of this type seems in fact to be widespread and progressing rapidly.

On the other hand, a social system directed towards greater humanisation has not been entirely neglected, as is clear from the well-established scientific field of human biology, and from various more or less related activities. However, virtually all such activities are still at a stage where engineering methodology is being used for the study of social phenomena.

Our constantly advancing civilised society is now living in the shadow of limited energy resources. It should be our future task to lead our social system towards a more skilful utilisation of energy through the development of resource-circulation systems and greater use of ''natural'' biological systems. For this it will be essential to cultivate understanding of the interdependence of energy and information, which is the basis of the environmental system shown in Fig. 16.6.

This is not at present taking place. Looking at technical systems of mechanisation, we see that so-called man–machine systems (even those based on scientific research into human behaviour) are leading more and more towards *non*-humanisation and an overall trend towards labour-saving. In this sense labour-saving systems, as well as earlier systems of social engineering, will surely become isolated, for as yet there is no established methodology to link mechanisation with humanisation. This we must develop from the concepts of human biology in the material (natural) world.

An ''adjustment technique'' able to resolve such a contradictory situation by scientific technology must be called Eco-technology. For ecology offers essential knowledge from nature to form an environmentally harmonious system for this purpose. Naturally-occurring events can be divided into animate and inanimate processes: such natural events have today become parts of a semi-natural or artificial system, and this results in confusion. The object of Eco-technology is to develop a scientific method for bringing about harmony or coexistence between the natural system and the artificial system, and this leads to the concept of a *United Eco-Mechanism*, as in Fig. 16.8.

Essential International Cooperation

Eco-Technology involves the much-discussed concept of ''appropriate technology'', a concept which originated in the *Intermediate Technology Development Group* (ITDG) founded in 1965 by the late Dr. E. F. Schumacher*. The ITDG was in fact set up to develop a so-called intermediate technology; to this end it has

*Once Economic Adviser to the British National Coal Board, but better known for his advocacy of the rather general thesis ''Small is beautiful''.

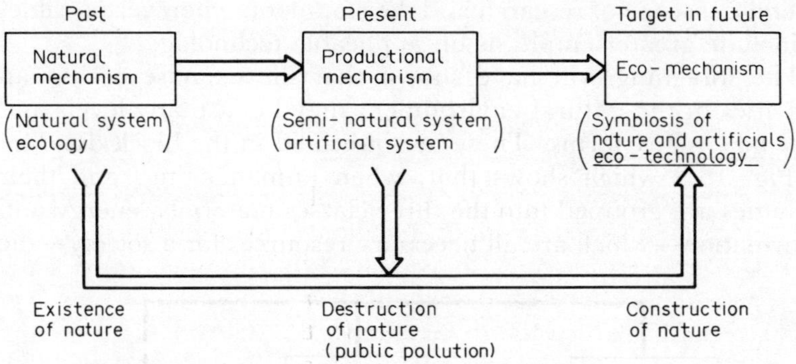

Fig. 16.8 Mechanism of past, present and future in environmental system

been working, in close cooperation with the United Nations and some twenty other organisations, to identify the roles which technology should play in such development.

Dr. Schumacher's work has had a great influence on the various organisations of the UN, resulting notably in investigations of how technology might best help the developing countries. African nations in particular have established development research groups in several specific fields (agriculture, civil engineering, resources development, etc.) whose work has proved most valuable.

Resulting from these activities, "applicable technology" (practical appropriate technology) has spread on a world-wide scale and its concepts have been adapted to resolve some of the problems of developed industrialised countries. Results achieved, whether in developing or developed countries, have amply shown the immense potential of technology in a great many fields; it has also become clear that a much wider dissemination of "technical know-how" is essential if the needs of all countries are to be met.

Among the developing countries there are some which already have a certain degree of technical background, and in these countries the low-cost, intermediate and "soft" techniques of applicable technology already exist. Moreover, since this type of technological evolution is normally accompanied by a renovation of social and cultural levels, it would seem most valuable to extend the scope of research and development, wherever possible, to include greater emphasis on applicable technology.

The foregoing will have shown that the exercise of human activities in the natural environment may be at many levels and may take many forms. These are indicated in the block diagram of Fig. 16.9 which shows that, when human beings and their activities are grouped into the three classes materials, energy and information—which are all necessary resources for a society—the

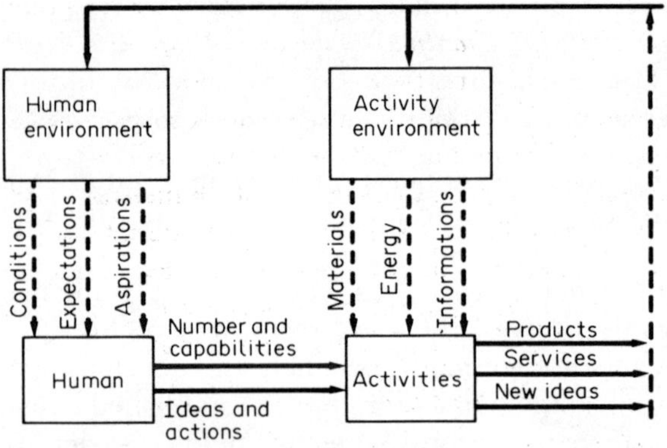

Fig. 16.9 Feedback loops in a system of human activities and environment

behaviour of the human beings is dependent on their conditions, expectations and aspirations, three factors which can evidently play a most important role*. The numbers of human beings involved and their potential for action (a quantitative measure of human resources) determines ability to transform physical resources in the three groups into the many and various kinds of products and services that human society requires. Some products and services will then be further synthesised to promote further (and better) human activities. This process is indicated by the feedback loop at the right in Fig. 16.9.

*If this is considered a statement of the obvious, it is nevertheless by no means always overtly recognised.

The "growth" of a nation means in general the increase of its products and services. Both these are clearly dependent *quantitatively* on human activity (more activity gives more products and services). The "development" of a nation, however, is influenced not only by its material resources but also by the social, cultural and economic conditions of its human society; the evaluation of "development" must therefore include a *qualitative* standard.

In selecting applicable technology, it is very important for a nation to decide on the relative importance to be given to these two distinct targets, *quantitative growth* or *qualitative development*, possibilities in each direction being evidently limited by the relation between the nation's material resources and material needs. This very important decision will of course affect the nation's material prosperity (in its most elementary—or material—sense) but it may have an even greater effect on whether the nation believes itself to be prosperous or not. This means that concepts of applicable technology, chosen for their practicability from social as well as material viewpoints, may well vary from one nation to another according to what each decides is its most important objective.

This in turn must of course affect the assessment of the appropriate technology for transfer to a not-yet-fully-developed nation. In practice and at present, although such technology transfers are made in many fields and in many ways, the above criterion (quantitative versus qualitative advance) is rarely considered. Nevertheless, there has for some time been a trend towards the transfer of information rather than materials—especially in advanced technical fields—and this trend seems generally advantageous, representing as it does the transfer of seed for local fertilisation rather than mature fruit with already determined characteristics. The inescapable conclusion is that education and training, carefully chosen with regard to regional conditions and requirements, are indispensable complements to (and may well be more important than) any material technology transfer.

The subject of applicable *techniques* (intellectual methodologies, logic, mathematics, conceptual models) has spread widely into

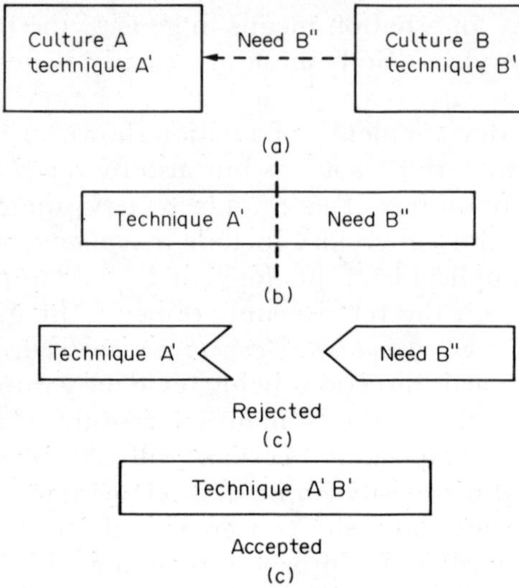

Fig. 16.10 Creation of applicable technique

domains far beyond technology in its normal sense, and is generally recognised as a matter of communication among different cultures. A transfer of this nature is represented in Fig. 16.10, where technique A′, developed within culture A, is to be combined with technique B′, an existing part of culture B, in order to meet a need B″ within or related to culture B. In other words, the technique A′ is to be introduced into the culture B to establish, in conjunction with the existing technique B′, a new technique A′B′.

This "applicable technique" cannot be achieved by the simple addition of technique A′ to technique B′ (i.e. by a simple technical transfer) but must take account of the two cultures involved, implying profound consideration and discussion prior to acceptance. Otherwise the new technique, no matter how skilful it may be and how great the need for it, is likely to be rejected. The seed of technique A′ will be culturally unacceptable for implantation in the soil of technique B′, so that the new applicable technique will not become established.

This demonstrates the very real difficulty of "modernisation through power". Technical transfer which neglects culture may provide a temporary form of modernisation, but in time this will be rejected by the cultural soil, possibly leading to a return of Luddism*.

The same difficulty can arise even without technical transfer. Many developing countries have promoted industrialisation with their own "cultural seeds", in an attempt to achieve rapid economic growth, but their social and cultural soils have rejected even these seeds and severe internal conflicts have resulted. In some cases, because of the shortage of "applicable" technical seeds and economic stock, which are indispensable for industrialisation, only partial modernisation has been achieved, bringing with it the new and severe problem of different levels of civilisation within the country itself. For example, in many cases where advanced centres for nuclear research, computers, or other highly technological activities have been established, and have yielded excellent results in their respective fields, large sections of the population are still trapped in repeating cycles of disease and poverty, and the misery and dissatisfaction of these people has become a significant factor impeding further modernisation. Although such situations are due in considerable degree to shortages of material and financial resources, and to very low levels of most existing technologies, there is no doubt that "underdeveloped" social and cultural structures are also responsible.

From this it must follow that, provided very careful attention is given to the social and cultural structures in a nation, including the representation of these structures in government, technical transfers and the introduction of applicable techniques should be possible even in conditions of material shortages, financial difficulties, and low levels of existing technologies. That is to say, it should be possible to overcome such obstacles by introducing a

*Deliberate destruction of machinery and technical equipment. After the *Luddites*, a band of English workmen who in the early nineteenth century tried to prevent industrialisation by wrecking factories and machinery.

system of effective government founded in popular support (the power of the masses). This has been demonstrated in certain African nations and in Japan.

On the other hand, however abundant its resources and extensive its lands, if a country has no philosophy of technical development and no concrete strategy based on such a philosophy, that country will not be able to eradicate poverty caused by its low standards of technology (as in the case of some other African countries). Even concerning technical development alone (the first essential step) if there is no national aim to promote it and no social demand for it, it will take place only sporadically and will stand very little chance of leading to techniques which could otherwise be the basis of collective national power. The essential strategy for progress must therefore be to pool individual techniques in an intensive and coordinated effort to generate a collective vitality throughout the country's entire industrial, social and cultural structures.

For developing countries in particular, such a strategy will necessarily involve adopting both the analytic theory of atomism and the collective theory of holism. Referring again to the experience of Japan, analytical consideration shows that development of technology has been the fruit of introduced, absorbed and improved techniques from other countries, and this process is often thought to be a pattern applicable to all the developing countries. That is, it is assumed that each novel technique and its related techniques will automatically promote the continuing development of technology. It is assumed that, when analysis of successful enterprises has identified the factors responsible for each success, the combination of those factors must automatically lead to new successful developments of technology.

However, if holistic social and cultural considerations are neglected, there are many unexpected pitfalls. For the synthesis of technical capacity, such a holistic basis is indispensable. Even though we may try to follow again, in today's developing countries, the methods which the former developing countries evolved through trial and error (even after now submitting those methods, where possible, to atomistic analysis) this cannot be regarded as a proper or valid substitute for a scenario based on a holistic approach to human history.

A fundamental dilemma facing all developing countries is whether they should seek primarily to introduce new technologies from the developed nations, or whether they should give their major attention to developing their own traditional technologies. Whichever path may be chosen, it will remain vitally important to start from an evaluation of *social purpose*, then selecting techniques which best respond to the most desirable purposes*.

Orders of priority in the development and employment of technology in developing countries should ideally be decided on the bases of the country's economic situation and aims, its material and financial resources, its geographic and climatic conditions, as well as its political and social structures and goals. If all these factors are taken fully into account, the applicable technology is likely to be virtually indistinguishable from the (most desirable) appropriate technology. Its pursuit is therefore particularly worthwhile, and Fig. 16.11 indicates schematically the desirable correlations between atomistically analysed culture

*Pitfalls will appear in this process also, notably because consensus over social purpose is more easily postulated than achieved.

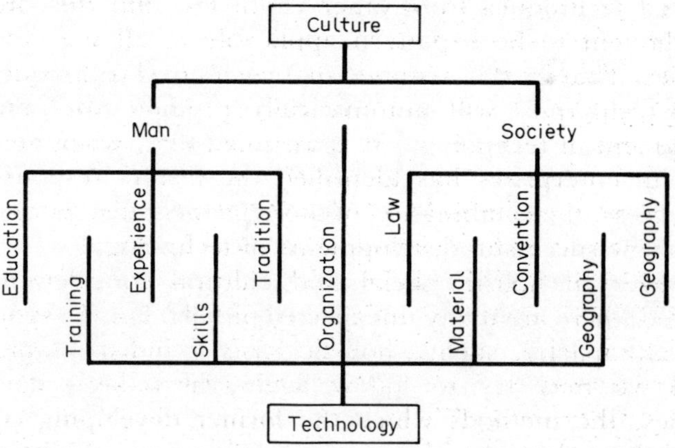

Fig. 16.11 Interdependence

and technology, and the holistic functions of culture as a symbiosis of man and society.

Fig. 16.11 in particular demonstrates that the central managing structure (''organisation'') can and should assume the role of joining the groups on both sides (''man'' and ''society''). This is the essential contribution of human intelligence to the entire holistic system, and it must be at the centre of any technological development strategy for the attainment of a national (or even an international) goal. It is both the conception and the function of the synthesis of culture with technology that is Eco-technology.

Bibliography

1. S. Aida, *Introduction to Eco-Technology* (Japanese), Gakken Publishing Co., 1973.
2. U. Pellegrini, *The Problem of Appropriate Technology*, Proceedings of IFAC Symposium on Criteria for Selecting Appropriate Technologies under different Cultural, Technical and Social Conditions, Italy, May 1979.
3. J. H. Armstrong, *Cultural Discontinuities—Their Effect on Technology and Development, ibid.*

Biographical Notes (in order of contributions)

Shuhei Aida is Professor of Systems Science at the University of Electro-Communications, Tokyo and a Lecturer in Bioengineering at the Tokyo Institute of Technology. He is also Director of the Honda Foundation. He has served in the National Biomedical Research Foundation in the USA and in the Laboratory of Cybernetics in Italy. He is Vice Chairman of the "Social Effects of Automation" project of IFAC (International Federation of Automatic Control) and in 1976 received the Pattern Recognition Society Award.

Publications include: *Introduction to Cybernetics; Robot; Introduction to Ecology; Science of Prediction; Systems Engineering.*

Bruce Adkins is an electrical engineer by profession, a science writer and an energy consultant. He was engaged in developing radar jamming equipment for the Royal Navy during the war and was a member of the Royal Naval Scientific Service after the war, until 1953, when he joined Elliott-Automation Ltd., holding several senior appointments on the information side.

In 1958 he became Head of the Public Relations and Information Service of what is now the OECD Nuclear Energy Group and since 1979 has been counsellor and consultant to the Organisation.

His publications include many technical and popular articles on radar, space travel, energy and world problems.

Harold W. Lawson, Jr. is Professor of Telecommunication and Computer Systems at the University of Linköping. He has held senior and visiting professorial appointments at the Royal Institute of Technology in Stockholm, the Polytechnic Institute of Brooklyn, the University of California in Irvine and in universities in Barcelona, Gothenburg, Stuttgart, Brooklyn and New York. He has also worked in the Standard Computer Corporation, International Business Machines Corporation and Sperry Univac. His primary research areas include computer architecture and the computer-aided design of large and very large scale integration circuits.

Eduardo R. Caianiello is Director of the Post-Graduate School of Cybernetics at the University of Salerno. From 1956 to 1972 he was Professor of Theoretical Physics at the University of Naples and until May 1977 was Founder and Director of the Laboratory of Cybernetics of CNR, Naples. He has also been a visiting professor and lecturer at many international universities and institutions in Rochester, USA, Princeton, Madras, Saclay, the Sorbonne, London, Chicago, Tokyo, the Soviet Academy of Sciences, etc.

His publications include *Combinatorics and Renormalization in Quantum Field Theory* and more than a hundred publications in various topics of theoretical physics and cybernetics.

TORGNY SEGERSTEDT is currently Professor Emeritus at the University of Uppsala, where previously he was Professor of Moral Philosophy from 1938 to 1947 and Rector Magnificus from 1955 to 1978. He is a member of the Swedish Academy, and of the Swedish Academies of Science, Letters and Engineering Sciences. He was Chairman of the Swedish Social Science Council from 1966 to 1975 and from 1966 to 1975 President of the Bank of Sweden Research Foundation.

His numerous publications are in the fields of sociology and philosophy.

ALAIN TOURAINE, after being in charge of research at CNRS from 1950 to 1957, became Director of Studies at l'Ecole Pratique des Hautes Etudes (VIe Section) and Founder and Director of the Laboratory for Industrial Sociology at l'Ecole Pratique des Hautes Etudes. He was also Director of the Department of Sociology in the Faculty of Letters at Paris-Nanterre from 1966 to 1969, and President of the Société Francaise de Sociologie from 1968 to 1970. He has held visiting professorships at Columbia, Montreal, UCLA, Chile and Sao Paulo. He is also Vice President of the International Association of Sociology.

Co-founder of the journal *Sociologie du travail,* he has published some sixteen books, edited seven and is the author of numerous articles.

HAROLD A. LINSTONE is Professor of Systems Science at Portland State University and the Director of its Futures Research Institute. He was formerly Senior Scientist at Hughes Aircraft Company and Associate Director of Corporate Development Planning at Lockheed Aircraft Corporation. His academic appointments include one as Adjunct Professor of Industrial and Systems Engineering at the University of Southern California from 1965 to 1970.

He is co-editor of the *Delphi Method, Technological Substitution*, and *Futures Research: New Directions*. He is also editor of *Technological Forecasting and Social Change.*

TOSHIRO TERANO is Professor in the Department of Control Engineering at Tokyo Institute of Technology and was formerly on the research staff of the Transportation Technical Research Institute for fifteen years, from 1947. During the academic year 1957/58 he was a visiting professor at MIT. Vice President of the Japan Association of Automatic Control Engineers and Vice Chairman of a technical committee in IFAC (International Federation of Automatic Control), he has awards from the Japan Society of Mechanical Engineers, the Society of Instrument and Control Engineers and the Association of Shipbuilding Engineers.

His numerous publications are in the fields of automatic control, nuclear engineering, security of process plant, artificial intelligence and man-machine communications, among other subjects.

REIKICHI SHIRANE is President of the Telecommunications Science Foundation of Tokyo. From 1966 to 1970 he was Secretary General of the Japan Technoeconomics Society, followed by an appointment as Managing Director, Data Communications Head Office, Nippon Telegraph & Telephone Public Corporation, 1970 to 1975.

His publications are *Operations Research, Technology Assessment*, and *Network Society.*

UMBERTO PELLEGRINI is Professor of Applied Electronics at the University of Applied

Electronics, University of Milan. He was Chairman of the National Committee for the "FAST–White book on job profile transformation as a consequence of evolution in technology" and Chairman of the National Committee for the preparation of the "FAST–White book of electronic industry in Italy" in 1976. His industrial appointments include General Manager of MONTEDEL (from 1969) and Corporate Advanced Technology Manager at Olivetti Company. From 1969 to 1973 he was President of ANIPLA, the National Scientific Association for Automation, and from 1975 he has been President of "FAST"—the Federation of Scientific and Technical Associations in Milan.

His publications include more than fifty scientific publications on digital electronics and three volumes of academic lectures on applied electronics.

Heinz Zemanek is Professor and IBM Fellow at the Technical University of Vienna. He is a Past President of IFIP (International Federation for Information Processing) and his main fields of activity are in information theory, computer voice-output on the Vocoder-principle and in the theory of programming languages. He has been involved in the development of a formal definition method for programming languages (known as the Vienna Method) applied to the formal definition of the programming language PL/I.

Ithiel de Sola Pool is the Ruth and Arthur Sloan Professor of Political Science at the Massachusetts Institute of Technology, where he has taught since 1963. Before that he taught at Stanford University and at Hobart College. He was a member of the Surgeon General's Scientific Advisory Committee on Television and Social Behavior, 1969 to 1971, is a Fellow of the American Academy of Arts and Sciences and was a Foreign Visiting Fellow of Churchill College, Cambridge, England.

His publications are principally in the fields of communications and public opinions and include *Talking Back: Citizen Feedback and Cable Technology; The People Look at Educational Television; The Social Impact of the Telephone* and *Handbook of Communication*, 1973. He has carried out studies on the impact of television viewing in politics and on audience research programmes.

Fred Margulies is Secretary of the Austrian Trade Union of Salaried, Clerical, Commercial and Technical Employees and a Lecturer at the University of Technology at Vienna. Since 1955 he has been active in the Trade Union movement, especially on automation problems, and has worked on problems concerned with the social impact of the computer and of technological change in general.

Since 1978 he has been Honorary Secretary of IFAC and, since 1976, vice chairman of a Technical Committee on Computers and Society in IFIP (International Federation for Information Processing).

R. Narasimhan is Director of NCSDCT (National Centre for Software Development and Computing Techniques) at the Tata Institute of Fundamental Research in Bombay.

He was President of the Computer Society of India from 1965 to 1969, a Fellow of the Indian Academy of Sciences since 1966 and a Fellow of the Indian National Science Academy since 1975. His principal field of interest is in applying systems theory concepts and techniques to the study of behaviour and the behavioural sciences and he has worked on models for visual perception and language behaviour.

His publications include papers on visual data processing, programming language theory and models of language behaviour.